Avian Orientation and Navigation

Avian Orientation and Navigation

Klaus Schmidt-Koenig

University of Tübingen, Tübingen, West Germany and
Duke University, Durham, North Carolina, USA

1979

Academic Press   London   New York   San Francisco

A Subsidiary of Harcourt Brace Jovanovich, Publishers

ACADEMIC PRESS INC. (LONDON) LTD
24/28 OVAL ROAD
LONDON N.W.1

U.S. Edition Published by
ACADEMIC PRESS INC.
111 FIFTH AVENUE
NEW YORK, NEW YORK 10003

British Library Cataloguing in Publication Data

Schmidt-Koenig, Klaus
   Avian orientation and navigation.
   1. Bird navigation
   I. Title
   598.2'1'8   QL698.8   79-50526

   ISBN 0-12-626550-x

Printed in Great Britain by
Clarke, Doble & Brendon Ltd, Plymouth and London

Preface

The mysteries of animal migration, and especially of bird migration, have intrigued man for centuries. Vanishing animal stocks and dwindling resources mean that more and more emphasis is placed on conservation. One prerequisite for successful conservation of free-living animals is a good understanding of the whereabouts of the animals to be protected. This includes knowledge of the means of orientation and navigation of animals that migrate. Research and the public availability of results are, therefore, essential elements of the environmental battle which is currently being waged.

In recent years several books on animal migration and orientation, covering birds as well as other animals, have been published (e.g. Orr, 1970; Schmidt-Koenig, 1975). The proceedings of three symposia (Adler, 1971; Galler et al., 1972; Schmidt-Koenig and Keeton, 1978) treated selected aspects from that general field. A series of major reviews specializing in current research on bird orientation has also appeared; Schmidt-Koenig (1965) was able to cover work on migratory birds as well as work on pigeons; Keeton (1974a) and Emlen (1975 have both published reviews of equal size, Keeton covering mostly homing pigeons and Emlen mostly migratory birds. There were very few books, however (the exceptions are Salomonsen, 1969 and Schüz, 1971) which were primarily about bird migration. Another example is Matthews (1968, 1971) but this was essentially propagating his sun arc navigation hypothesis and cannot therefore be considered impartial.

This decade has seen an explosion of ideas, experiments, data and interpretations in the field of bird orientation. The published literature has grown to such a volume that reviews now have to specialize on certain topics such as compasses (Waltschko, 1973), and more comprehensive accounts necessarily attain book size. Even the author of a book of reasonable format must, nowadays, be selective to some extent. Everything published on or related to bird orientation cannot be considered unless the author is willing to produce an encyclopaedia. The present book is not intended to be an encyclopaedia, but rather an attempt at a fairly comprehensive review, emphasizing the most important and most promising aspects relevant to a solution of the various unsolved problems of bird orientation and navigation.

Several attempts have been made to define and classify orientational phenomena such as directional orientation, homing or navigation in birds. We are most concerned here with "homing" i.e., finding some goal. Griffin (1952, 1955) proposed a classification of orientation into three types. Type 1 was a search for familiar landmarks to be used for piloting home. Type 2 was the ability to head in a given compass direction, by itself insufficient for homing, unless taking up a certain direction always leads home or to familiar landmarks. Type 3 was true navigation, the ability to orient towards a goal by means other than the utilizing of familiar landmarks.

Schmidt-Koenig (1965, 1970a) modified Griffin's definition and subclassi-fied true navigation into (a) reverse displacement navigation, and (b) bi-co-ordinate navigation. Since the word displacement may be considered to involve only passive transport, the term "route reversal" may be more appropriate (Mittelstaedt; *see* Schmidt-Koenig, 1970a). Theoretically, route reversal may be accomplished e.g., by inertial navigation systems, as proposed by Barlow (1964) (Chapter VI, Section D). The sensory system would continuously sum changes from the home setting. In a "one-step" opera-tion, the displaced bird need only to take up a heading that diminishes the precession or change from the home value. "Two-step" operations could be visualized as transforming the inertially obtained information regarding direction towards home into a compass direction. As the second step, a compass would be utilized for actual homing. Instead of inertial information during the outward journey, information from some compass e.g., the sun compass or magnetic compass may be "read" and processed by the bird. Bico-ordinate navigation may be accomplished on the basis of astronomical co-ordinates as proposed by Matthews (1953b) (Chapter VI, Section B) or by geophysical co-ordinates as proposed by Yeagley (1947) (Chapter VI, Section A). These would represent "one-step" operations in which the animal moves in the direction that resets the co-ordinate values to the goal values. By analogy to two-step inertial systems a compass could also be used in a two-step fashion, which would be a verification of Kramer's concept of the map and compass.

These subclassifications still stand but the term navigation must be reconsidered. "Familiar landmarks", excluded from true navigation, were usually assumed to be visual landmarks, and homing from "unfamiliar sites" has been postulated as one prerequisite for using the term true navigation. With Papi's findings that olfactory cues are either carried by winds to the home position and sampled at home, or recognized and processed during displacement or at the release site (Chapter V, Section E), previous definitions may have to be reconsidered. Even when a pigeon has been displaced to a location where it has never been before it may be familiar to the bird through an odour originating there. This odour or any other site-specific stimulus other than visual landmarks may also be considered a familiar mark. Homing, which is perfectly possible on the basis of olfactory (or other) information from new sites would, therefore not qualify as navigation.

Although the question of how birds find their goals is unsettled, I feel that navigation may at the moment be defined as the capacity to find or to establish reference to a goal. This definition has moved away from the classical definition of navigation as deriving one's position from astronomical data only.

Acknowledgements

I am indebted to a number of colleagues for their helpful discussions and also for the additional information they provided, including J. Kiepenheuer, Tübingen, W. T. Keeton, Ithaca, Ch. Walcott, Stony Brook and W. Wiltschko, Frankfurt. Particular thanks should go to D. R. Griffin, New York, W. T. Keeton, Ithaca and J. Kiepenheuer, Tübingen, for critically reading the manuscript and for the many suggestions they made to reduce the number of shortcomings and mistakes. While writing his book I held research grants from the Deutsche Forschungsgemeinschaft whose support is gratefully acknowledged. I would also like to thank B. Schwolow, H. Tabel, I. Rindfleisch and K. Bok for patiently typing the manuscript and drawing the figures.

Contents

I. Introduction to migration

Bird migration is one of the most fascinating natural phenomena, and it is still a highly mysterious one, especially if we ask how birds find their ways and destinations, i.e. how they navigate. Before engaging in a discussion of the problems of orientation and navigation, I would like to consider, firstly, what birds accomplish in migration.

More than half a century of bird banding has revealed the whereabouts of many species of birds during the course of the year. Numerous species, such as the English sparrow (*Passer domesticus*), remain in the same area during the entire year, i.e. they do not migrate. In other species, however, things may be more complicated. Some populations of a species may stay in their breeding ranges all year round while other populations of the same species migrate regularly between a breeding range and a winter range. An example of this kind is the fox sparrow (*Passerella iliaca*), which lives on the Pacific coast of Alaska and Canada. Figure 1 shows the geographical distribution of six populations of this fringillid. The southern-most population (6) does not migrate, while populations 1–5, living north of population 6, migrate and winter south of population 6.

There are numerous well-established examples of birds that migrate along fairly straight lines between summer and winter ranges not too far apart, and sometimes without having to cross major geographical barriers such as large bodies of water or mountains. Recoveries of Swedish wood pigeons (*Columba palumbus*) as documented by Rendahl (1965), may serve as an example (Fig. 2). The Baltic populations of the starling (*Sturnus vulgaris*) would represent another well-documented example discussed in a different context in Chapter VI, Section B. One may safely conclude that these short to medium distances are covered in straight lines. Many migratory birds are known, however, to change course one or even more times on their way to or from their winter range. The migratory routes of the eastern and western European populations of the white stork (*Ciconia ciconia*) around the Mediterranean (Fig. 3) would represent a

Fig. 1. Breeding ranges (solid lines) and winter ranges (broken lines) of fox sparrows. *Passerella iliaca* (1) *unalaschcensis*, (2) *insularis*, (3) *sinuosa*, (4) *annectens* (5) *townsendi*, (6) *fuliginosa*. After Schüz (1971).

Fig. 2. Foreign recoveries of wood pigeons (*Columba palumbus*) banded in Sweden. From Rendahl (1965).

fairly simple case of this category. Detouring the Mediterranean east or west both populations show a major turn towards a more southerly direction, the western populations at Gibraltar and the eastern population near the gulf of Iskenderun.

Several major turns have been described by Kullenberg (1946) and Storr (1958) regarding the populations of the arctic tern (*Sterna paradisaea*) inhabiting arctic Canada and Greenland (Fig. 4). The birds cross the Atlantic heading east and then turn south at the western Atlantic coast of Europe, some to follow the western coastline of Africa, some to cross the Atlantic again for the east coast of South America to spend the winter at the southern tip of South America or Africa or further south in the Atlantic Ocean around Antarctica. The arctic tern also holds the long distance record for all avian migrants, one leg of the journey extending for at least 10 000 km.

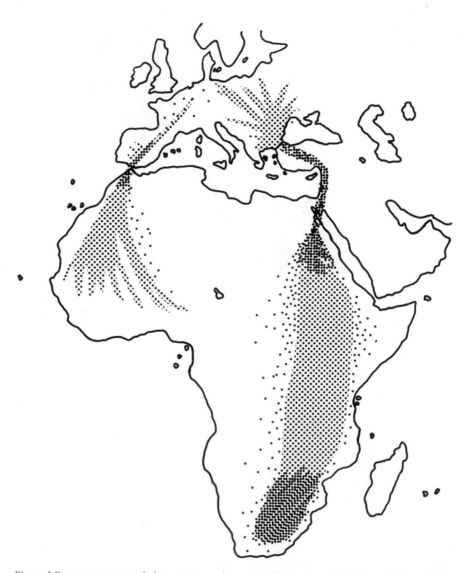

Fig. 3. Migratory routes of the eastern and western European populations of the white stork (*Ciconia ciconia*). Both populations fly around the Mediterranean west or east, and most birds have to change their headings from a southwesterly or southeasterly direction to a southerly direction. After Schüz (1971).

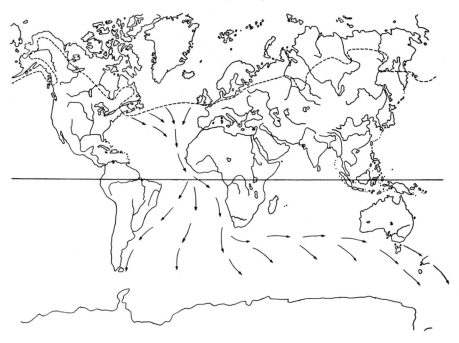

Fig. 4. Autumn migratory routes of the Canadian and Greenland populations of the arctic tern (*Sterna paradisaea*). After Storr (1958).

The arctic terns probably reverse their autumn migration in spring, as do most other migrants, but there is evidence from banding and observational data that birds take different routes in spring and autumn. The golden plover (*Pluvialis dominica dominica*) may serve as an example (Fig. 5). Golden plovers breed in most parts of arctic Canada, and in the autumn migration are considered to cross the Atlantic non-stop from Labrador and Nova Scotia to the lesser Antilles and northeastern South America, and pass on to the winter range in southern Brazil, Uruguay and northern Argentina. Spring migration seems to take a more westerly route by way of northwestern South America, Central America, the Gulf of Mexico and up the Mississippi flyway into Canada.

Another example would be the slender-billed shearwater (*Puffinus tenuirostris*), which leads us to the question of migration in the southern hemisphere. As shown in Fig. 6, the slender-billed shearwater breeds on parts of the southeastern coast, on some islands of Australia and on Tasmania. Its annual migration describes a circle clockwise around the Pacific touching the coasts of Japan, Kamtchatka, the Aleutians, Alaska and the North American Pacific coast to southern California (Serventy, 1953; Marshall, 1956).

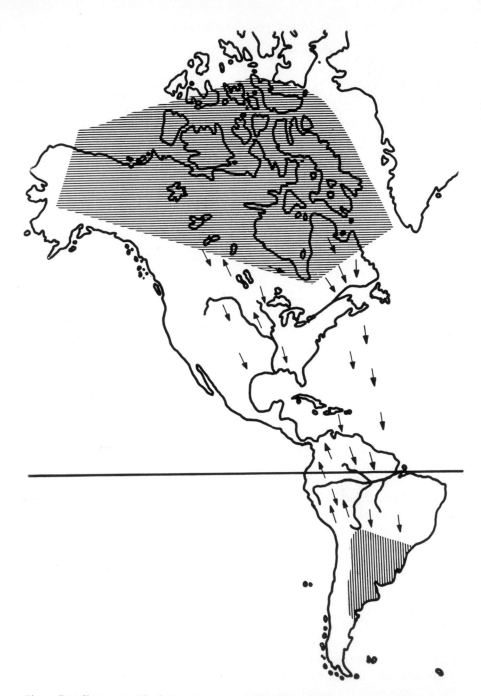

Fig. 5. Breeding range (shaded), winter range (shaded) and approximate autumn and spring migration routes of the golden plover (*Pluvialis dominica dominica*). Modified from Salomonsen (1969) and from Schüz (1971).

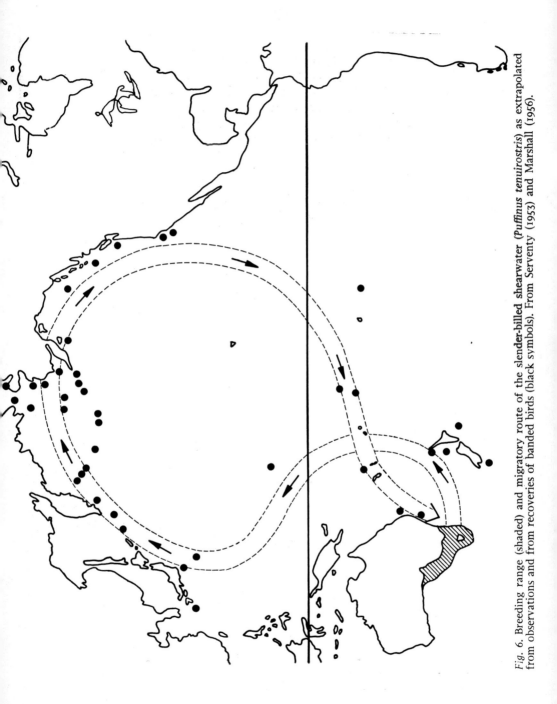

Fig. 6. Breeding range (shaded) and migratory route of the slender-billed shearwater (*Puffinus tenuirostris*) as extrapolated from observations and from recoveries of banded birds (black symbols). From Serventy (1953) and Marshall (1956).

There is bird migration in the southern hemisphere, or from the southern to the northern hemisphere, as well as vice versa. However, since there is less land in the southern hemisphere birds are less numerous, as all birds require land at least for breeding. Also, large parts of the land of the southern hemisphere are less civilized and bird migration has attracted less attention in the southern hemisphere, as it is harder to watch bird migration in a barren environment.

Nevertheless, many species of birds breeding in, e.g. southern South America, southern Africa, Australia and New Zealand migrate roughly north in their autumn and south in their spring. There are species such as, for example, the carmine bee-eater (*Merops nubicoides*) breeding in parts of southern Africa, which move just a few hundred kilometres northward without reaching the equator (Chapin, 1932). Some cuckoos from Australia and New Zealand migrate considerably longer distances north, to winter on Polynesian islands (Fig. 7).

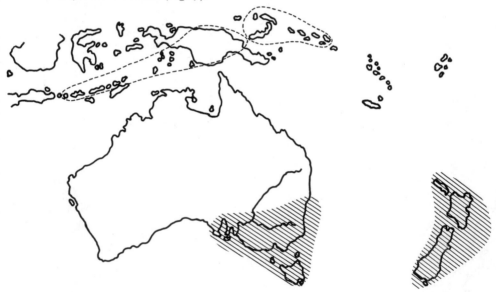

Fig. 7. Breeding range (shaded) and winter range (broken lines) of the bronce cuckoo (*Chalcites lucidus plagosus*) from Australia and *Chalcites lucidus lucidus* from New Zealand. After Fell (1947) and Dorst (1962).

Data on migration in Asia and equatorial latitudes, which are less well known than data from Europe and North America, have been accumulated by McLure (1974). It must also be remembered that many large scale avian movements in Australia, Africa or South America are related to droughts, rainy seasons, floods, fires etc. and the subsequent lack or abundance of food, and are not therefore the same as seasonal migration.

II. Migration recording techniques

A. Field-glass observation

The last decade has seen a revolution in the methods of recording bird migration. Recording by field-glass observation has been carried to its limits, for example by Kramer (1931), who discovered high-altitude migration beyond the visibility of the naked eye and by Lowery (1951) and Lowery and Newman (1966), who watched nocturnal migrants against the disc of the moon. Gauthreaux (1972) introduced yet another variation that made nocturnal field-glass observation independent of the moon and clouds. Gauthreaux pointed a strong light into the night sky and recorded the direction of migrants moving through the light beam. Bellrose (1971) has carried direct observation to its present extreme by recording nocturnal migrants from a small aeroplane equipped with powerful lights. All these methods of direct observation provide partial estimates of the volume and direction of bird migration. The species involved usually remain unidentified and, except in the data of Bellrose, only migration near the ground is sampled. Finally, listening to flight calls permits identification of some species, but estimation of the volume and direction of migration is extremely difficult.

B. Radar

A new era of recording bird migration began with the use of radar. Radar recording has the advantage of being largely independent of weather conditions because it works at night as well as during the day and it is unaffected by clouds. It has the disadvantage, however, of being expensive and requiring expert maintenance.

There are two different types of radar systems, i.e. surveillance radar and tracking radar. The surveillance radar is the type used at airports: its continuously rotating antenna covers a certain radius around its location like a street light. With additional techniques to read the screen, e.g. time-lapse photography, it reveals whether or not there is much migratory activity and the general direction and speed of movement. Figure 8 presents a normal photo of a radar screen showing extensive migration of chaffinches (*Fringilla coelebs*) over the English Channel, south-eastern England and the coast of France: direction of movements is not revealed. In contrast to this "snap shot" figure, Fig. 9 shows a two-minute time-exposure photograph of a radar screen. The general direction of movement, which is north-west to south-east is now revealed. Data of

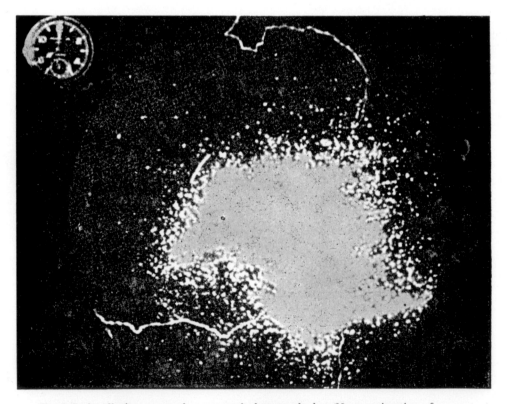

Fig. 8. Radar display as seen by a normal photograph shot. Heavy migration of chaffinches over southeastern England, the English Channel and the coast of France. The coastlines of Norfolk (upper right) and of western Sussex (lower left) can be distinguished. From Lack and Eastwood (1962).

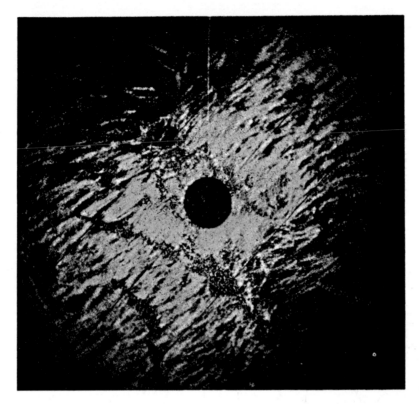

Fig. 9. Radar display as seen by time-exposure photography. Streaks indicate migratory movement north-east–south-west. From Gering (1963).

this kind may be analysed in relation to meteorological events. Figure 10, for example, represents the result of such an analysis (Lack and Eastwood, 1962): lapwings (*Vanellus vanellus*) seem to fly to the south across the English Channel away from snow showers (the hatched areas) moving in from the north.

Many details essential for the analysis of bird migration are, however, not revealed by surveillance radar, e.g. number and identification of birds, the heading of the individual or flock, altitude etc. cannot be obtained. Thus, surveillance radar provides a more qualitative and general picture of bird migration.

A vast literature has been accumulated and names like Lack, Sutter, Eastwood and Drury are associated with radar ornithology; and Eastwood (1967) summarized the state of radar ornithology in a sizeable book, which also gave detailed information on radar operation. After much initial discussion and many apparently conflicting observations most experts

Fig. 10. Radar evidence in relation to weather factors. The arrows indicate movements of lapwings (*Vanellus vanellus*) moving south across the English Channel away from snow showers (hatched) moving in from the north. From Lack and Eastwood (1962).

now agree on the following findings: birds migrate in a directed fashion not only under clear or partially clouded skies, but also under total overcast. Instances of disorientation are rarely seen on radar but are always associated with fog, total overcast or both. But total overcast as such does not suffice for disorientation, as only when presumed alternative guiding factors are also eliminated does disorientation result. Birds may, at least to some extent, compensate for wind drift and they may change their directions in flight. In contrast to this ability to maintain a direction there is as yet no clear-cut evidence as to whether the appropriate migratory direction is determined during take-off under overcast. Recent findings by several authors (Emlen and Demong, 1978; Able, 1978; Gauthreaux, 1978) point to the important role of sunset as a possible directional cue for migration during the night.

More recently a different type of radar has been used, i.e. the tracking radar. In contrast to the surveillance radar, the tracking radar acts like the beam of a torch. Its narrow beam detects and follows a single target, or a small group of birds, and provides precise quantitative data from this one target, such as distance from the radar site, direction of movement, altitude, speed, change of direction, change of speed etc., all of it printed out if necessary. The track over ground can be plotted automatically on a map. If wind data are recorded by releasing and tracking balloons the heading of the migrant may also be calculated. Figure 11 presents the radar tracks of a number of birds migrating across Bermuda as published by Williams *et al.* (1972). In addition, some radar sets are equipped to record, and plot some equivalent of the wing-beat pattern of the birds. An experienced tracker can identify at least some species or families. Figure 12 shows the "radar signature" of three birds: a mistle thrush (*Turdus viscivorus*), a chaffinch (*Fringilla coelebs*) and a white wagtail (*Motacilla alba*), the identity of which was confirmed visually with a telescope during the daytime (Bruderer and Steidinger, 1972). Identification of birds migrating only at night is difficult and has not always been accomplished. It has, however, been confirmed by this technique that swifts (*Apus apus*) continue flying overnight in their breeding range in Europe (Bruderer and Weitnauer, 1972).

Returning to migration, tracking radar has provided valuable details of the behaviour of migrants and the influence of, e.g. altitude, speed and wind. Figure 13 presents data obtained by Bruderer and Steidinger (1972) indicating that on one spring night birds seemed to migrate in three altitude levels, the lower from 400–1100 m, the middle from 1400–1700 m and the upper from 1900–2100 m; wind and bird directions and speed of the lower and the upper level being clearly different. Wind and bird directions do not, however, seem to be directly related. Birds migrate in all sorts of winds including tailwind, headwind and crosswinds, although

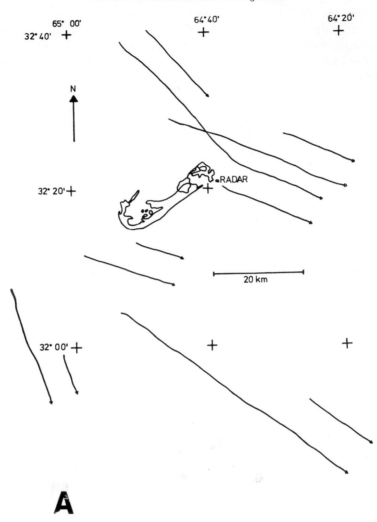

65° 00'
32° 40'
64° 40'
64° 20'
N
32° 20'
RADAR
20 km
32° 00'

A

Fig. 11. Radar tracks of birds migrating across Bermuda Island. A, birds moving south-east; B, birds moving south-west. From Williams *et al.* (1972).

they apparently prefer tailwinds. Whether or not birds compensate for wind drift, i.e. whether or not they advance their headings sufficiently to offset the effect of crosswinds, has long been discussed in the literature. There is evidence that birds may compensate and there are other examples in which they did not compensate for wind drift. No clear answer has, so far, been presented as to which circumstances or factors may be responsible for the alternative behaviour.

The bulk of migration at north-temperate latitudes has been shown to

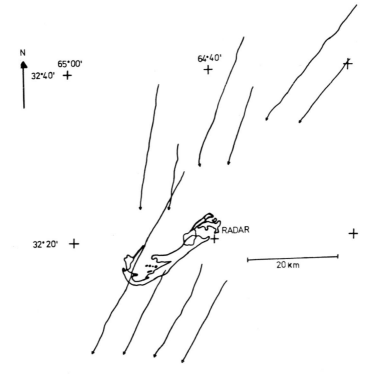

N

65°00'

32°40' +

64°40' +

32°20' +

RADAR

20 km

+

B

occur during only a few days and nights with favourable weather conditions. Dense autumn migration quite predictably occurs, in the New England area, after the passage of a cold front on the east side of a high pressure area. In spring, dense migration is observed on the west side of a high
pressure area ahead of an approaching depression (e.g. Drury and Keith,
1962; Nisbet and Drury, 1968; Richardson, 1971, 1972). Both conditions
provide tailwinds for the appropriate direction of migration, southerly in
autumn and northerly in spring. Tailwinds would certainly lower the
energy cost of flight, but migration by tailwind only would be difficult

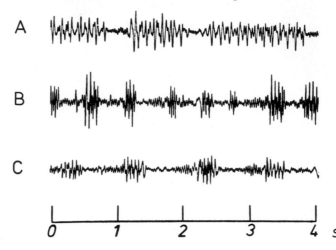

Fig. 12. "Radar signature" of, A, a mistle thrush (*Turdus viscivorus*), B, a chaffinch (*Fringilla coelebs*), C, a white wagtail (*Motacilla alba*), that had been identified visually. The wing-beat frequencies and the periods of flapping and pausing are species-specific. From Bruderer and Steidinger (1972).

to accomplish. Migration in winds other than tailwinds is common, though wind may be an important cue for many migrants. It is possible that downwind flights regardless of direction is the predominant mode in some portions of migration, i.e. in the southern United States, as suggested by Gauthreaux (1972) and Able (1972, 1973, 1974, 1978). Environmental conditions, birds and findings may, however, be different elsewhere (e.g. see Alerstam, 1976; Bruderer and Winkler, 1976; Bruderer, 1977, 1978).

There is some evidence that regularly blowing winds such as the trade winds have been incorporated by migrants into their migration strategy. Large numbers (though probably not the majority) of small passerine birds depart in autumn from the coast of Nova Scotia and New England heading over the Atlantic in a southeasterly direction (Drury and Keith, 1962; Drury and Nisbet, 1964; Richardson, 1972), maintaining a southeasterly direction at least within radar range (up to 150 km). Some are not seen on land again before they arrive in the Caribbean Islands and South America. According to radar evidence (Williams *et al.*, 1972) large numbers of migrants pass non-stop over Bermuda, usually (though not always) heading south-east. This bearing would lead them, if they continued, southeastward between Africa and South America. Perhaps many die but a number large enough to guarantee the survival of the species or at least that fraction heading south-east, reaches the Caribbean Islands and South America, riding or drifting with the trade winds from the north-east. Williams *et al.* (1974, 1977) and Williams and Williams

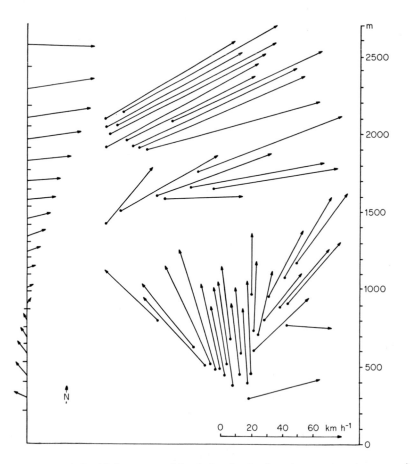

Fig. 13. Tracks of the birds near Zürich, Switzerland, plotted as vectors (arrows with specific length according to the scale in km h^{-1} and direction, with north up) at various altitudes above ground (right-hand ordinate). The bird's altitude is indicated by the dot at the origin of the vector. Wind direction and speed (measured by ceiling balloons) as a function of altitude, are shown on the left-hand ordinate, also as vectors in the same scale. From Bruderer and Steidinger (1972).

(1978) have attempted to survey autumn migration on a large scale by organizing simultaneous observations with a network of nine radar stations—partly surveillance and partly tracking radar types—extending from Halifax, Nova Scotia to Miami, Florida, a boat in the Atlantic between New England and Bermuda and on several of the Lesser Antilles (Fig. 14). There are some indications that many birds spend 80 hours or more flying non-stop from Nova Scotia (or correspondingly less when leaving from the eastern United States) to reach the Lesser Antilles or South America. There is also evidence for considerable migration along the east coast of North America.

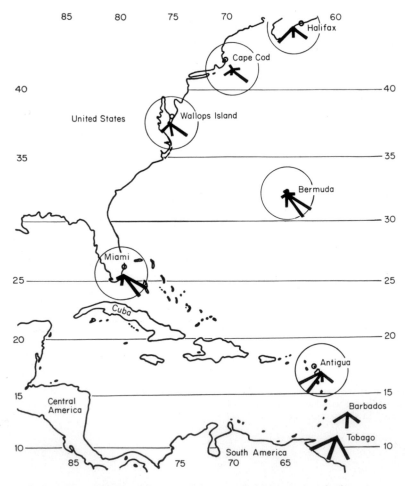

Fig. 14. Eight radar stations used by Williams *et al.* (1974) to watch the transatlantic migration in the autumn of 1973. From Williams *et al.* (1974).

For the student of navigation the transatlantic route(s) are, of course, more exciting than the internal routes. Though tracking radar provides more information on single migrants than surveillance radar, the observer is still dependent upon chance to find migrating birds and he often does not know which species he is recording, the origin and destination of the bird etc. The observer cannot interfere experimentally to find out more about the birds' mechanism of orientation. Emlen (1974) succeeded in overcoming at least some of these problems, by placing freshly caught migrants, i.e. identified species, into cardboard boxes and releasing the box suspended from a balloon near a tracking radar. The radar locked on the bird in the box, the bird was released at a pre-set altitude and was tracked on its way. This technique enabled Emlen to release birds with magnets attached to them to disturb their possible magnetic orientation, or to release birds after their internal clock had been reset to test the use of the sun compass etc. Emlen and Demong (1978) provided initial evidence that night migrating white-throated sparrows (*Zonotrichia albicollis*) may use the glow of sunset as a directional cue. When released from the balloon under clear starry skies, the sparrows rapidly selected the appropriate migratory direction and maintained altitude. When released under total overcast excluding stellar cues from view, northerly directions were still selected but in a much less pronounced fashion.

C. Radio tracking

Still another approach to studying the orientation behaviour of migrants is to use radio telemetry or radio tracking. Modern technology has reduced the size and weight of the necessary components to such a degree that microtransmitters of just a few grams can be made, small enough to be carried by a medium-sized passerine such as, e.g. a thrush. Cochran (1972) deserves a considerable part of the credit for pioneering the development of this technique. Resting migrants are caught, a transmitter is attached—usually glued—to the bird, and the bird is released. The experimenter then has to listen to the transmitter signal reproduced by his receiver and to wait for the bird to take off. In Cochran's experiments birds sometimes took off soon after release to continue their migration, but a bird sometimes spent hours before moving on, testing the experimenter's alertness and uselessly drawing energy from the battery of the bird's transmitter.

The receiving antenna may be stationary, mounted on a truck or attached to an aeroplane. A mobile or airborne antenna has the advantage

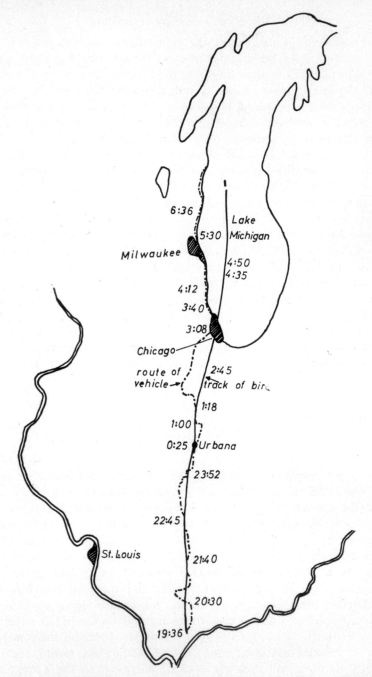

6:36

Lake
5:30 Michigan

Milwaukee

4:50
4:35

4:12

3:40

3:08

Chicago

route of
vehicle →

2:45

track of bird

1:18

1:00

0:25 Urbana

23:52

22:45

St. Louis

21:40

20:30

19:36

Fig. 15. Spring migratory flight of a veery (*Hylocichla fuscescens*) from southern Illinois out over Lake Michigan. Approximate track of the bird (solid line with the time of day indicated) and route of the car carrying the antenna (dashed line). From Cochran (1972).

of increasing the otherwise very limited range of radio tracking (a few miles only). By moving along with the bird in a car or plane, Cochran was able to track single passerines over several hundred kilometres and waterfowl such as whistling swans (*Olor columbianus*) over more than 2000 km. Figure 15 presents an example of a veery (*Hylocichla fuscescens*) caught, tagged and released during spring migration in southern Illinois and tracked for 11 hours and severel hundred kilometres until the signal was lost over Michigan.

Many more tracks have been obtained and published (e.g. Graber, 1965; Cochran *et al.*, 1967; Cochran, 1972). Mostly, the birds continued their migration in a more or less straight fashion, some flying through thunder-storms or fog. More information, especially on longer flights to or from the winter range, may be obtained from automated tracking via satellites. Radio tracking has also been employed extensively in homing experiments with homing pigeons. These will be reviewed in Chapter V (Section D).

III. Laboratory experiments

Migration always involves considerable distances and time spans. This makes observational work—not to mention experimental interaction—extremely difficult and is responsible, at least in part, for the slow progress in solving the age-old problems of bird migration. Birds can, however, be caged and brought to the laboratory and at least part of the navigation process may be studied under laboratory conditions. The advantage of laboratory work is that a bird can be exposed to controlled conditions such as an altered magnetic field, altered photoperiod, artificial celestial constellations etc., which are impossible to control or even to monitor in a free-flying migrant. The disadvantage of laboratory studies is that the whole process of navigation cannot be reproduced in the laboratory.

Laboratory experiments can be subdivided into those based on the bird's migratory restlessness, and those using training or conditioning procedures which are not necessarily dependent on the bird being in a physiological state of migration.

A. Basic sensory capabilities

Bird migration and pigeon homing are largely unexplained. Even the sensory modalities involved have remained, for the most part, unknown, though many sensory modalities have been suggested as responsible for, or at least involved in, migration and homing. Investigations of sensory capacities possibly involved in orientation and navigation have lagged behind the evidence of orientational and navigational accomplishments. As late as 1965 (Schmidt–Koenig, 1965), almost nothing from the field of sensory physiology relevant to orientation and navigation could be reviewed. Fortunately, this situation has changed significantly.

1. Sensitivity to ambient pressure changes

Employing the technique of cardiac conditioning (shown in Fig. 20) Kreithen and Keeton (1974a) tested the ability of 12 homing pigeons to detect air pressure changes in a constant-environment chamber. Ten of the birds responded regularly to pressure changes at intervals of 5 s duration. Figure 16 shows the response pattern of one of the birds which was

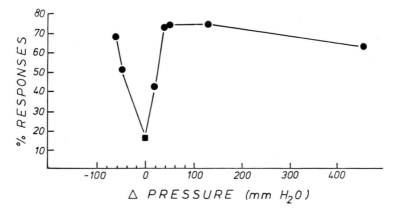

Fig. 16. Response to the atmospheric pressure changes of one bird in cardiac conditioning experiments. The solid square core designates control experiments without pressure changes. From Kreithen and Keeton (1974a).

tested over a fairly wide range. Including the results from other experiments with modified pressure changes, the authors were able to conclude that homing pigeons are apparently able to detect atmospheric pressure changes of the order of 10 mm H_2O up or down within 10 s without sharp thresholds, but with some evidence of response to pressure changes of as little as 1 mm H_2O. Delius and Emmerton (1978) also investigated pressure sensitivity in pigeons and fully confirmed Kreithen and Keeton's (1974a) findings. Delius and Emmerton (1978) suspect that Vitali's organ, located just below the middle ear lining, is the sensory organ responsible for the pressure sensitivity, but definite evidence has not yet been produced.

A pressure change of 10 mm H_2O is equivalent to a change of approximately 10 m altitude. While perhaps not directly related to orientation such a physiological altimeter could be very useful for maintaining altitude in flight without visual contact with the ground. Radar studies have shown that migrating birds maintain altitude within 20 m, even inside clouds (Griffin, 1969, 1972). There is also a remote possibility that a physiological "barometer" might be used for pressure-pattern navigation as discussed by Kreithen and Keeton (1974a). Detection of pressure changes would, of course, be useful in order to anticipate certain meteorological changes. This would be important for the migrant not only to help it avoid bad weather, but also to predict and take advantage of following winds that may greatly reduce the energy requirement of a migratory flight. If migratory routes cross inhospitable ground such as deserts or oceans the ability to predict the weather and any favourable winds correctly, may be of vital importance and high selective value in the evolution of bird migration. There are some indications that migratory restlessness of caged migrants is correlated with barometric pressure (Masher and Stolt, 1961; Muller, 1972) and there is evidence that much migration occurs on selected days and nights, which are to some extent predictable from meteorological indicators such as pressure changes (e.g. Lack, 1960; Drury and Keith, 1962; Nisbet and Drury, 1968; Richardson, 1971, 1972).

2. Hearing—infrasound

Acoustic waves of frequencies below 10 Hz are called infrasound because man cannot hear them; they are below the human threshold of hearing. Yodlowski *et al.* (1977) and Kreithen (1978) tested the lower threshold of pigeons in cardiac conditioning experiments very similar to those shown in Fig. 20. As shown in Fig. 17, birds unexpectedly responded to sounds as low as 0·06 Hz. In a search for the location of the receptors sensitive to infrasound, the cochlea and lagena were surgically removed in some birds. In others both columellas were bisected so that conduction of sound between the tympanum and the oval window was impossible. No further response to infrasound was recorded in any of the experimental birds. This result suggests that the receptors for infrasound are located in the inner ear and that infrasound is truly "heard" and not perceived as pressure changes by ordinary pressure sensors.

Infrasound can travel over distances of several thousand kilometres without much attenuation. There are many natural sources of infrasound, some of which are of high amplitude. Stationary sources include surf and mountains (especially coastal mountains) with the winds forming a sort of infrasound flute, and moving sources include thunderstorms, magnetic storms and aurorae, low pressure areas etc. Rockets and supersonic aeroplanes would be examples of man-made sources of infrasound.

It is as yet unknown whether birds can discriminate between the various

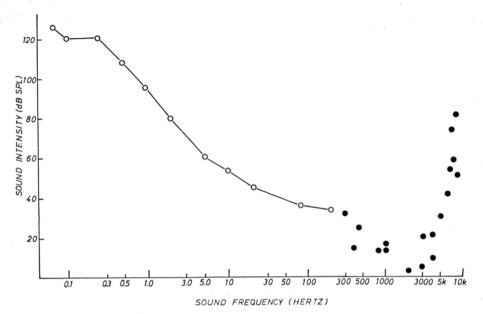

Fig. 17. Sensitivity of a homing pigeon to infrasound obtained in cardiac conditioning experiments. Each open score symbolizes the 50% threshold at that frequency (abscissa). Solid scores are from other laboratory experiments. From Kreithen (1978).

frequencies of atmospheric infrasound, especially so-called "pseudosound" which is caused by intense fluctuations in atmospheric pressure at infrasonic frequencies, or whether they can locate the sources or at least the direction of them. If so, they could theoretically be used for triangulation even when moving slowly. Birds are known to allow for slow movements such as the apparent movement of the sun in sun compass orientation. Directional identification of a source of infrasound by the same mechanism as is used to locate sources of audible sound would require ears more than one kilometre apart because of the extremely long wavelength. Location identification could, however, be accomplished by using the Doppler shift. Kreithen (1978) calculated a possible frequency shift of 14% for a pigeon flying towards or away from the sound source at a speed of 20 m s^{-1}. This could enable the bird to find the direction of the sound source. In a preliminary study by Kreithen and Quine, mentioned in Kreithen (1978), pigeons have responded in laboratory conditioning experiments to shifts in sound frequencies of less than 5% at 5 Hz. Further research is needed to clarify the matter.

3. Olfaction

Most birds are usually considered to have a poor sense of smell, but the recent findings by Papi and his co-workers (Chapter V, Section E; and Chapter VI, Section E) mean that this capacity should be reassessed. The laboratory evidence was last reviewed by Wenzel (1971a). According to Bang and Cobb (1968) and Bang (1971) the size of the olfactory bulb as a percentage of the largest diameter of the forebrain ("olfactory bulb index"), taken as a correlate of olfactory capacities, attains the largest values among Apterygiformes (34·0), Procellariiformes (29·0), Podicipediformes (24·5), Caprimulgiformes (23·8) and so on, and the smallest values among Piciformes (10·0), Passeriformes (9·7) and Psittaciformes (8·0). The size of the pigeon olfactory bulb (20·0; Bang, pers. comm.) is in the intermediate range between these. Wenzel (1971b) implanted electrodes into the olfactory bulbs and recorded changes in electrical activity upon presentation of odorous stimuli in seven species of birds such as the black-footed albatross (*Diamedea nigripes*), black-vented shearwater (*Puffinus puffinus opithomelas*), turkey vulture (*Cathartes aura*), chickens (*Gallus gallus*), mallards (*Anas platyrhynchos*), raven (*Corvus corax*) and pigeon (*Columba livia*), i.e. from the upper, intermediate and lower olfactory bulb classes. Details of the experiments on pigeons were published by Sieck and Wenzel (1969). The odorants used were exclusively synthetic chemicals such as amylacetate, pyridine, methylsalicylate and trimethylpentane. The thresholds were measured to be at concentrations of the order of 10^{-8}–10^{-10} mol ml^{-1} of odorant in air and electrical activity was found to change with concentration. Bilateral olfactory nerve bisection abolished all responses to the odorants. Comparable results have been reported by Shibuya and Tucker (1967) and Wenzel and Sieck (1972) and some were found to be similar to those recorded from macrosmatic reptilian and mammalian species (Tucker, 1965). Perception of naturally occurring odours has not been tested.

Results from conditioning experiments using either heart rate or respiration or both, as indicators of perception of odours, confirms the evidence discussed above, at·least for pigeons (Henton *et al.*, 1966; Michelsen, 1959; Shumake *et al.*, 1969; Wenzel, 1967) or for mallards (Whitten, 1971) using motor restlessness as a purely behavioural response. The absolute thresholds for amlyacetate were found to be 0·16–0·73% of vapour saturation (Henton, 1969) and the difference thresholds to be 0·57 and 0·71 (Shumake *et al.*, 1969).

Other species tested by cardiac or respiratory conditioning include the bob-white (*Colinus virginianus*), canary (*Serinus canaris*), Manx shearwater

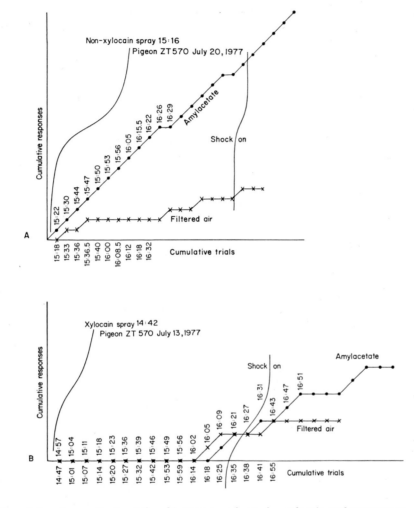

Fig. 18. Local anaesthesia of the olfactory membrane by xylocain and response to 4·5% amyl acetate in a cardiac conditioning experiment (ordinate). (A) The response persisted after application of control spray lacking xylocain. (B) There is no response for 60–90 min (abscissa) after application of xylocain. From Schmidt-Koenig and Phillips (1978).

(*Puffinus puffinus*), chicken, turkey vulture (*Cathartes aura*), Humboldt penguin (*Spheniscus humboldti*) and the kiwi (*Apteryx australis*), and all these produced positive evidence for perception of at least some odours, the only exception being the raven (*Corvus corax*), which did not produce any responses. Again standard laboratory chemicals only were used.

In those birds with a high olfactory bulb index use of olfaction for foraging has been suggested or experimentally demonstrated, e.g. in vultures (Staeger, 1964) and kiwis (Wenzel, 1971b)—though older evidence was clearly negative (Strong, 1911). More evidence from field work with wild birds is discussed at the end of Chapter IV, Section E.

The pigeon, with its intermediate olfactory bulb index, has been brought to the centre of attention by Papi as possibly using olfactory cues for homing. In conjunction with homing experiments with pigeons whose olfactory membrane was anaesthetized by xylocain (Chapter V, Section E) the effective elimination of olfaction by local anaesthesia was tested in cardiac conditioning experiments by Schmidt–Koenig and Phillips (1978). In an olfactory chamber the response to 4·5% amylacetate persisted after application of a non-xylocain control spray provided by the manufacturer (Fig. 18A). The response was eliminated for 60–90 min upon application of xylocain spray reduced from human otolaryngeal dosage (Fig. 18B). The relevant homing experiments are discussed in Chapter V, Section E.

Schmidt–Koenig and Phillips (1978) failed to condition the same pigeons to discriminate between air samples collected out of doors (which presumably contained those odorous substances the birds use in homing, according to Papi), and filtered air. Thus all the evidence from field work is indirect, and no direct evidence from laboratory conditioning or electrophysiological experiments is available. Different methodological approaches may well produce positive results but at the present time laboratory work does not directly support the hypothesis of olfactory homing in pigeons. Moreover, there is considerable laboratory evidence that surgical manipulations of parts of the olfactory system such as lesions or ablations of the olfactory bulb have distinct effects on non-olfactory functions such as visual learning and discrimination, including orientation tasks and motivational aspects, especially in pigeons (Wenzel and Salzman, 1968) and mammals (e.g. Lindley, 1930; Phillips, 1969; Marks *et al.*, 1971; Alberts and Friedman, 1972). Such aspects may be relevant to the rather puzzling results obtained in experiments on olfactory pigeon homing.

4. Magnetoreception

The division of this chapter into "basic sensory capabilities" and "integrated sensory capabilities" cannot be strictly maintained. The bulk of work on the use of the geomagnetic field is work on magnetic compass orientation and therefore clearly falls into the section on "integrated sensory capabilities". Magnetic sensitivity such as is involved in the work of e.g. Bookman (1978) may turn out to fall into the section on "basic sensory capabilities" as does Leask's (1977, 1978) model. All work on the use of magnetic cues is discussed in the section on the magnetic compass; see also Chapter III, Section B, 4 for the basic points on magnetoreception.

5. Vision

(a) Visual spectrum

Various aspects of vision may be relevant to orientation and navigation though the traditionally assumed central role of vision, at least for homing in pigeons, can no longer be maintained.

Earlier experimental attempts to determine thresholds of sensitivity for the pigeon, starling (*Sturnus vulgaris*), and American robin (*Turdus migratorius*) (Blough, 1956, 1957; Adler and Dalland, 1959; Adler, 1963) seemed to indicate that night probably appears darker to these birds than to man.

Previous conclusions regarding the limits of the visual spectrum in birds have turned out to be false. Kreithen (1978), again employing cardiac conditioning, extended the tests to include ultraviolet light. Previous investigators had all discontinued investigation of shorter wavelengths where the glass used in the apparatus sets a barrier, but Kreithen, using Pyrex equipment, recorded responses to the ultraviolet range with maximum sensitivity at 360 nm and at 325 nm (Fig. 19). If older results are reanalysed in the light of this the published curves do appear to show a bend just before the glass barrier. This was, at that time, either taken as a chance event or an artifact and not as the indication of a return of sensitivity at shorter wavelengths. The visual spectrum of homing pigeons at shorter wavelengths now appears to be roughly the same as that of insects. If confirmed these findings would constitute a major discovery. It is as yet unknown how this sensory capability is utilized by birds. Schlichte (1973) found no indication of an effect caused by ultraviolet-absorbing contact lenses in homing experiments with pigeons. There is some preliminary

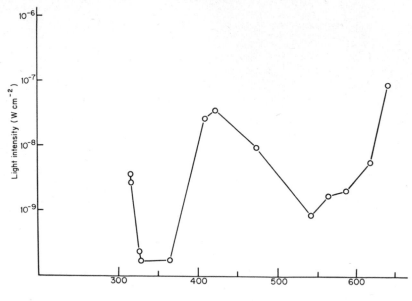

Fig. 19. Sensitivity of a pigeon to visible light and to ultraviolet light as revealed in cardiac conditioning experiments. Each score symbolizes the 50% threshold at the wavelength (abscissa). From Kreithen (1978).

thought that the perception of the plane of polarized light is mostly accomplished in this range of spectrum. This will be discussed below in Section (b).

(b) *Plain of polarized light*

It has been known for some time (e.g. Chard, 1939; Catania, 1961, 1964; Galifret, 1968) that the retina of the pigeon is divided into an upper or peripheral part used for near vision and characterized by red oil droplets, and the central and lower part for distant vision characterized by yellow oil droplets. In earlier experiments Montgomery and Heinemann (1952) had been unable to train pigeons to discriminate between the plane of polarized light on pecking discs in the red field of the retina. Kreithen and Keeton (1974b) presented the polarized light at a distance of 1·9 m in the yellow field of the retina, and the birds then responded. The same chamber used by Kreithen in barometric pressure and spectral sensitivity experiments was now modified for testing detection of the plane of polarized light (Kreithen and Keeton, 1974b; Kreithen, 1978). The experimental apparatus is shown in Fig. 20. The bird is trained to respond to rotation of the polarizing filter and responds with an increase of its heart rate (as in all experiments of this general type). A response curve

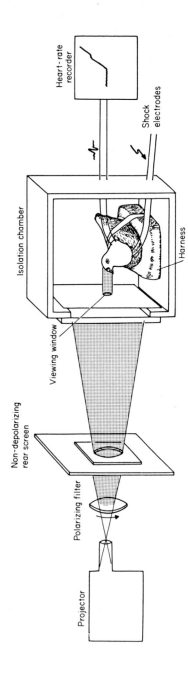

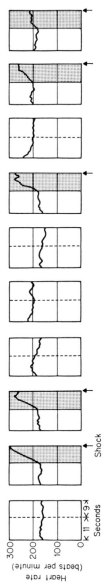

Fig. 20. Diagrammatic sketch of cardiac conditioning experiments of a pigeon in an isolation chamber as used by Kreithen; the experimental set-up modified for testing perception of linearly polarized light. The pigeon is strapped down inside the sound-proof chamber. The projector was switched on at random intervals. Heart rate recordings of five tests with rotation and of five control tests without rotation are shown in the rectangular diagrams below. Adapted from Keeton (1974b).

B*

(Fig. 21) clearly shows the bird's ability to perceive the rotation of the plane of linearly polarized light. Only four out of 12 birds which were used for this experiment could be trained successfully, which was considered a puzzling result. More recent evidence of sensitivity to ultraviolet light may resolve this puzzle. Polarization of the blue sky is strongest in the ultraviolet range and it is possible that the birds' detection of polarized light is predominantly localized in that range. Further experiments will clarify this issue. Delius et al. (1976) and Delius and Emmerton (1978) confirmed the detection of the plane of polarized light by pigeons using a different approach from that of Kreithen; they presented the stimulus overhead, and it was therefore, also perceived by the lower part of the retina.

Blue sky light is linearly polarized, the plane of polarization being a function of the position of the sun. It has been shown in bees (von Frisch, 1968) that sky polarization can be used for sun compass orientation substituting the sun if the sun is obscured by clouds, mountains or is below the horizon. It remains to be demonstrated how pigeons or other birds use their corresponding capacities.

(c) Visual acuity

The question of the acuity of vision leads on to the Section on integrated sensory capacities, as it has an intermediate position. A high degree of visual acuity in birds has been inferred from observations. Birds of prey or insectivores locate prey, and birds in general locate, especially flying predators, much better than man, although motion and not basic visual acuity appears to be the most important factor. Confirmation of the assumed high acuity of vision was extrapolated from anatomical evidence about the retina and the dioptric apparatus (Matthews, 1955, 1968). In recent experiments of modified operant conditioning Blough (1971) found thresholds of monocular distant vision of homing pigeons to be in the range of 1·94′–4·00′. These values are comparable to some findings or estimates of others for pigeons (e.g. Hamilton and Goldstein, 1933; Walls, 1963) or other birds such as swifts, buntings (genus Emberiza) or thrushes (e.g. Donner, 1951; Oehme, 1962) though different methods make direct comparisons extremely difficult. Blough (1971) tested human subjects with the same technique as she tested pigeons and found better visual acuity in man (0·79′ and 0·82′) than in pigeons. The fovea of the pigeon is considerably shallower than that of many other birds. Deepness of fovea and the acuity of vision are related, so one could therefore predict that pigeons do not rank among those with excellent visual acuity.

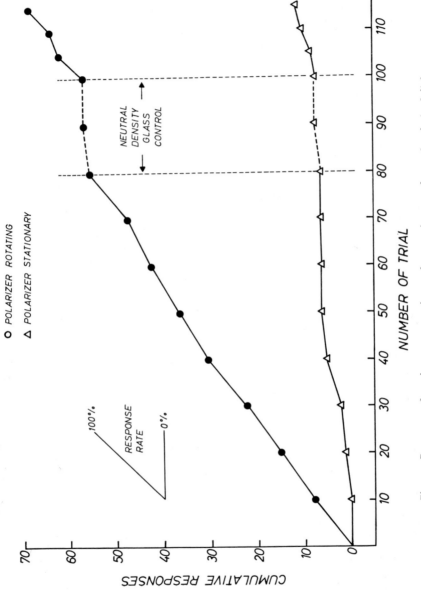

Fig. 21. Response of a pigeon to rotating and to stationary plane of polarized light. From Kreithen and Keeton (1974).

B. Integrated sensory capacities

1. Image vision

Visual acuity is certainly involved in image vision and this, in turn, is in all probability responsible for recognizing familiar landmarks. With cardiac conditioning experiments Sonnberg (1972) used an apparatus very similar to that used by Schlichte (1971, 1973) which is shown diagrammatically in Fig. 22. Sonnberg investigated to what extent pigeons visually recognize firstly, objects as such and secondly, structures in the intermediate vicinity of the loft.

If they are trained to recognize, from a few metres distance, an arbitrarily selected red pole of 5 cm diameter and 80–110 cm length the birds were able to discriminate it from similar test objects if the two were presented simultaneously. If only test objects were presented the birds seemed to have a poor recollection of the training object and also reacted to poles different in colour or in length. Training and testing situations were varied and are difficult to summarize in a few sentences. Sonnberg found indications, however, that the spatial position of training and testing objects were of some importance. In other experiments Sonnberg's pigeons showed that they had a rather precise visual recollection of the immediate vicinity and the spatial relationship of the loft entrance within the range of 25 m. These experiments and results are rather complex and difficult to evaluate. They certainly need to be confirmed and the questions which arise should be investigated, possibly using a modified approach and certainly a location more suitable than the loft at Göttingen. This loft was housed in the roof-top of a six-storey building with space to test the birds' visual memory of the vicinity limited to a 15 × 15 m platform and the ridges of the roof, including the wings of the building. The same is true of Sonnberg's attempts to obtain a response towards the loft at the limits and outside the area of direct visibility of the loft. No positive result could be obtained at distances of 2·5–16 km.

Of basic relevance and importance to the homing experiments with frosted lenses (Chapter V, Section D), are those conditioning experiments in which Schlichte (1971, 1973) and Sonnberg (1972) tested the reduction of image vision caused by frosted lenses. Both found that pigeons recognized (through frosted lenses) such simple test objects as a red pole of 5 cm diameter or such complex structures as the loft entrance when 1 or 2 m away but they showed no reaction when it was 6 m or more away.

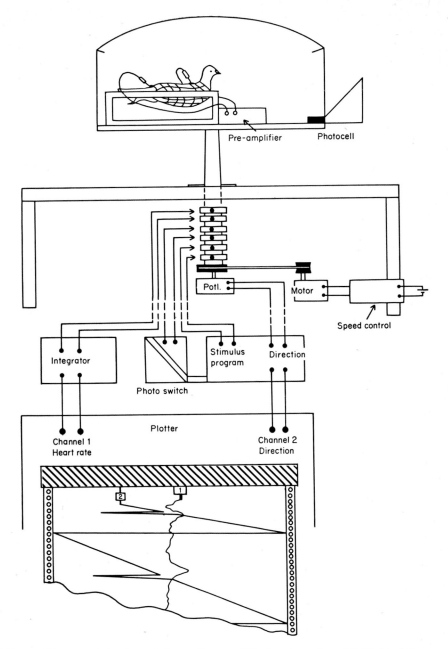

Fig. 22. Diagrammatic view of the cardiac conditioning apparatus of Schlichte (1971, 1973). The model given here was equipped with a photocell for sun training. Similar designs were used by Sonnberg (1972).

A new attempt to assign the pecten oculi a function in orientation has recently been made by Pettigrew (1978). The nutritive role of the pecten as demonstrated by Wingstrand and Munk (1965) still does not account for the heavy pigmentation, the distinct structure and, particularly, the precise geometric orientation within the eye. Measurements and observations on a scale model by Pettigrew indicated that the pecten casts a precisely defined shadow on the retina, which in relation to the retinal image of the sun, would facilitate some solar measurements. From the three theoretically possible kinds of solar measurements, i.e. sun compass, sundial or sextant, the sun compass function would be the one that can be derived directly from the pecten's shadow, giving the sun's azimuth. If additional information were used the shadow could be transformed into the time of day. Latitude could be determined (i.e. sextant function) if the bird measured the trajectory of the shadow tip across the retina and the position of the horizon, though the accuracy of this is questionable since the system is much less sensitive to changes in sun altitude than to changes in sun azimuth. There is as yet no experimental evidence in this area.

As discussed in Chapter IV, the sun compass as one system of directional information is well established in laboratory and field experiments. In contrast, the use of celestial information for navigation is not supported by experimental evidence (Chapter VI, Sections B, C).

2. The sun compass

In a pioneering study Kramer and St. Paul (1950a; Kramer, 1951) succeeded in training starlings in a circular cage to look for food in a certain compass direction. All visual landmarks were excluded by a screen and only the sky and the sun were visible. The birds maintained their training direction throughout the day. Kramer concluded that they allowed for the sun's apparent movement: he had discovered the sun compass. Observations of migratorily restless starlings (Kramar, 1950a) had preceded these experiments: a starling was kept in a cylindrical cage, also excluding any view of landmarks. The restless bird headed west-north-west. If the sun was deflected by a mirror (an old experiment introduced by Santschi, 1911), the bird's directional choice was deflected by the corresponding angle.

The subsequent training experiments (mentioned above) made experiments independent of the relatively short seasons in which the bird is in the physiological state of migratory restlessness, i.e. "co-operative" in showing its directed spontaneous migratory activity. The same capabilities were found in similar training cages in homing pigeons and meadow larks (*Sturnella neglecta*) (Kramer and Riese, 1952; St. Paul, 1956), and was later confirmed by others, with modified training procedures. The interaction between chronometry and the sun's apparent movement was particularly illuminated in experiments carried out by Hoffmann (1954), who demonstrated that the starling's chronometer or "internal clock" could be reset or shifted experimentally. The number of hours of shifting—in Hoffmann's original experiment 6 h counterclockwise—resulted in a predictable deviation of the bird's directional orientation in the training apparatus. The bird's clock is easily reset if the bird is confined in a light-proof room in artificial photoperiods, e.g. 6 h behind or ahead of the local photoperiod. In experiments with homing pigeons Schmidt-Koenig (1958, 1960, 1961) tested the effect of clock-shifts of 6 h clockwise, 6 h counter-clockwise and 12 h (to a reversed photoperiod). The results are given in Fig. 23. Schmidt-Koenig also demonstrated that a shift of 6 h clockwise or counter-clockwise takes four days to attain full effect, i.e. a 90° directional shift following 6 h, and a 180° directional shift following 12 h after resetting. The birds did not pay noticable attention to sun altitude even if expected sun altitude (e.g. near the horizon at 6 am) was grossly different from that actually seen (e.g. culminating at 12 noon), or if moving up rather than down.

After the sun compass had been firmly established experiments were extended to answer a number of relevant questions:

(*a*) how do birds cope with the latitudinal variability of the sun, i.e. in high northern latitudes and under equatorial and transequatorial conditions?

(*b*) how accurate is the sun compass?

Starlings (Hoffmann, 1959) and homing pigeons (Schmidt-Koenig, 1963a), after sun compass directional training at medium northern latitudes, have been tested in their training cage upon displacement north of the Arctic circle (starlings from Wilhelmshaven, Germany, 53°30′ N. to Abisco, Sweden, 68°22′ N.; pigeons from Durham, North Carolina, USA 36°00′ N. to Barrow, Alaska, 71°20′ N.). Both allowed, by and large, for the local sun during that part of the day that was also day in their home latitude. The starlings were also well-oriented under the midnight sun. The pigeons' sun compass orientation "at night" was somewhat less clear-cut, the details not being of sufficient interest to warrant a discussion here (c.f. Schmidt-Koenig, 1963a). Sun compass orientation on trans-equatorial displacement has been investigated in the same species by the same authors (Hoffmann, unpublished; Schmidt-Koenig, 1963b). Unfortunately, the starling is not a transequatorial migrant and the pigeon does not migrate at all. Both responded alike: they referred to the sun of the southern hemisphere—rising in the east, culminating in the north

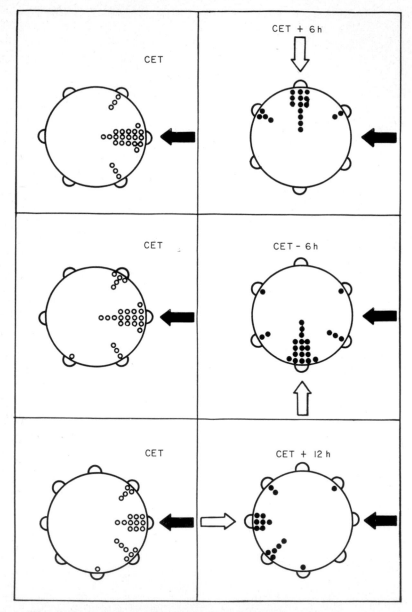

Fig. 23. The result of clock resetting on the directional choice of homing pigeons in training experiments. The cylindrical training cage is symbolized with six or eight choice points at the periphery. Each open dot symbolizes one choice (peck) of a pigeon under control conditions (natural day; left column). Each black dot symbolizes one peck of the pigeon under experimental clock shifts as indicated (right column). Black arrows indicate the direction expected under the experimental conditions of clock-shift. CET=Central European time. After Schmidt-Koenig (1960).

and setting the west—as if it were the sun at home, i.e. they were dis-
orientated. Attempts by Schmidt-Koenig (unpublished data) to repeat the
experiment with a passerine transequatorial migrant, the bobolink
(*Dolichonyx oryzivorus*), failed, in that the birds did not ever reach
South America. The experimental birds either died or escaped after training in Durham, North Carolina, had been established.

Attempts to establish the accuracy of the sun compass have not been
overwhelmingly successful. One technical drawback has been the wide
scatter which is always observed. The "classical" food-rewarded directional training as introduced by Kramer was later modified in many
ways. Figure 24 presents one of the modifications (Schmidt-Koenig,

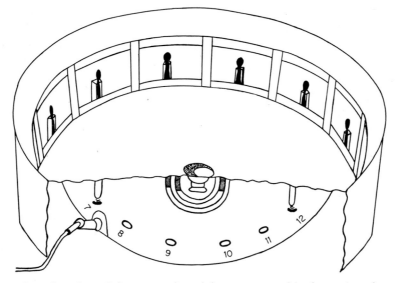

Fig. 24. A section view of the automatic training cage as used in the arctic and
transequatorial sun compass experiments by Schmidt-Koenig (1963a, b). Six of twelve
pecking discs (black) are shown at the periphery of the upper platform. A micro-
switch was activated with each peck at the disc. The electrical impulse passed
through brushes (two, to contact points 7 and 12 as shown) subtending from the
upper rotatable platfom to contacts 7–12 in the lower, fixed platform. After pecking,
the bird had to return to the food cup inserted in the centre of the upper platform,
which was operated automatically if the bird's choice was correct (rings and brushes
shown). Netting covering top and sides is not shown.

1963b): direct observation was replaced by automatic recording and rewarding, and the experimenter remained invisible. Food was not presented in the training direction at the periphery but in the centre of the
arena so that the bird had to return to the centre and start anew from
there for each choice. The number of choice points was considerably increased from the original 4 up to 24 (McDonald, 1973; Schmidt-Koenig

and Schleidt, unpublished). The diameter of the arena was also increased
and the training automated. Although the disadvantages of grouping (a
certain number of statistically discrete choice points as compared to
statistically continuous choice possibilities) was considerably reduced by
increased numbers of choice points, scatter of choices remained so
high that reasonable limits for the accuracy of the sun compass could
not be worked out. Even the question of the accuracy of the clock as one
integral component of the sun compass could not be satisfactorily
answered. As judged from other experiments using the period length of
locomotor activity, e.g. in homing pigeons (Miselis and Walcott, 1970), or
the onset of locomotive activity, as a "hand of the clock" in a variety
of vertebrates, the accuracy of the clock is in the range of, at best, several
minutes, the flying squirrel (*Glaucomys volans*) representing the best per-
forming animal (DeCoursey, 1962). Slightly more successful were the
attempts by Meyer (1964) and McDonald (1972) introducing operant conditioning methods. McDonald's apparatus is shown in Fig. 25: the bird
is trained to peck at key A to rotate the table, and its own long axis, in
small steps until the training angle, e.g. 15° counter-clockwise from the
sun, is reached. The bird then has to peck key B to get a food reward,

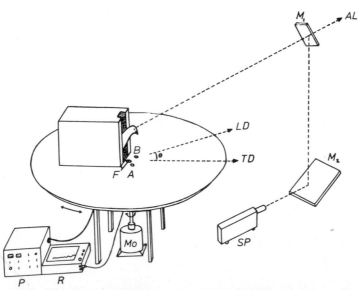

Fig. 25. McDonald's (1972, 1973) apparatus for testing the pigeon's accuracy in
measuring sun azimuth and sun altitude. The bird is fixed in the box and operates
keys A and B. F designates the food cup for rewarding. The table can be rotated by
a motor (Mo). The control unit (C) controls training and critical tests and the results
are recorded (R). In this arrangement an "artificial sun" is simulated by a projector
(PR) and presented to the bird by way of two mirrors (M₁, M₂). The angle α is the
training angle between the direction of the "sun" (LD) and the training direction (TD).
From McDonald (1972).

and the table returns to its original zero position for a new start. If, upon reaching the training angle, the bird continues to peck key A, the table rotates to the zero position without a reward. With this technique the bird trains itself automatically without any human interference. The bird's angular accuracy can be read from the recording paper. The pigeon's performance was quite disappointing: though fixed and sitting, and not exposed to the additional shaking of flying, the sun compass accuracy was between 3·4° and 5·1° one-sided, i.e. ±3·4° and ±5·1° two-sided. The accuracy for sun altitude relevant to sun navigation (Chapter VI, Section B) was between 8° and 11° one sided i.e. ±8° and ±11° two-sided.

A sun compass accuracy of between ±3·4° and ±5·1° is not as high as one would expect. However, computer simulations of homing flights (McDonald, unpublished data) suggested that a sun compass accuracy similar to that demonstrated in his training experiments, would be sufficient for success, and achieve the speeds actually recorded in homing experiments if the birds correct *en route* with sufficient frequency. This frequency is unknown. Above all, however, McDonald found that in his experiments the birds performed their sun azimuth (or sun altitude) training by measuring shadows rather than the sun directly (McDonald, 1972, 1973). McDonald (1972) calculated examples that demonstrated that measuring shadows may be of considerable advantage to birds by enlarging sun movement by a factor of up to almost six. It remained unknown, however, whether or not homing pigeons or other birds use shadows out of doors. Moreover, laboratory results from sitting birds cannot be extrapolated to flying birds. How accurately flying birds measure sun azimuth or other astronomical variables remains obscure. The question of how the sun compass is established ontogenetically has been tackled in homing experiments and is, therefore, discussed in Chapter V, Section B, 1.

3. The star compass

This field of study started with Kramer's discovery that the migratory restlessness of caged song birds is directed rather than random. Most observations were made during the day and some at night. Kramer continued with training experiments on sun compass orientation. Sauer and Sauer (1955) related the nocturnal directionality, upon confirming it in warblers (genus *Sylvia*), to the stars. The warblers were oriented under the natural starry sky as well as under a planetarium sky, and the direction roughly agreed with the ancestral migratory direction specific for the actual season, i.e. southerly in autumn and northerly in spring. In Sauer's experiments with warblers, data have been recorded by direct observation. The experimenter watched the birds from below and noted or dictated the direction of the bird and the time the bird fluttered in a certain direction. The fallacies of this technique yielding qualitative data, the small samples collected from only very few birds on which Sauer's interpretations were based, led to the rejection of Sauer's far-reaching conclusion that the warblers used stars, even in a planetarium, for bicoordinate navigation (for an explanation see Chapter VI, Section C).

Other recording techniques were developed yielding more quantitative and reliable data. Emlen and Emlen (1966) introduced the "funnel technique" for star orientation experiments with indigo buntings (*Passerina cyanea*) (Fig. 26). The footprint data can be converted into vector diagrams (Fig. 27) in three ways. Blotting paper is sectioned and the degree of blackening can be judged according to a standard scale by the experimenter; measured photometrically or the ink extracted from the blotting paper and measured. Emlen has usually used the first method.

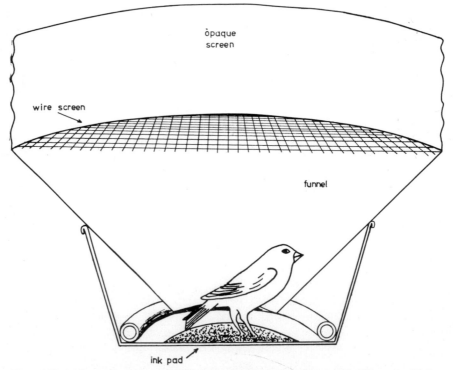

Fig. 26. Experimental cage as introduced by Emlen and Emlen (1966). When the bird attempts to leave the cage ink footprints are produced on the blotting paper lining the funnel. See also Fig. 27.

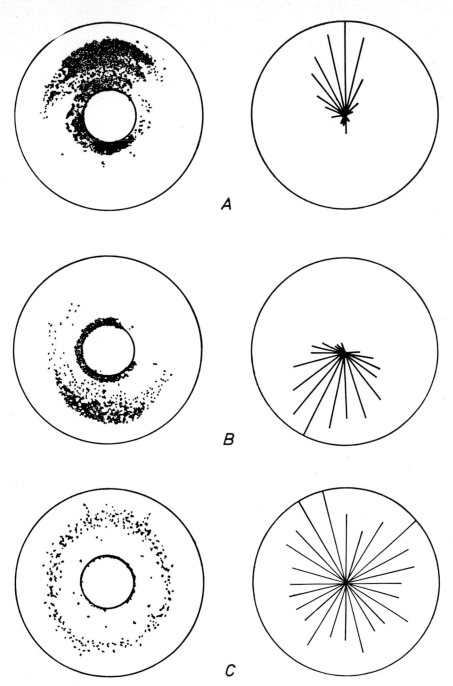

Fig. 27. A–C. Footprint raw data (left) as recorded by the funnel technique (Fig. 26) and their transcription into vector diagrams (right).
(A) Well-directed north-northwesterly orientation;
(B) Well-directed southerly orientation;
(C) Little or no directionality.

In his experiments with indigo buntings Emlen (1967a,b) first demonstrated that the directionality of the birds under the natural and under a stationary planetarium sky was the same, and that the birds followed a shift of the planetarium sky (Fig. 28). Upon manipulating the birds' photoperiod in such a way that experimental birds were in a physiological state of autumn migration at the time of spring migration, when tested under a spring planetarium sky the experimental birds headed south while control birds, unmanipulated and in the physiological condition of spring migration, headed north.

Attempts to find constellations crucial for orientation by selectively eliminating, e.g. the Big Dipper, Cassiopeia, Polaris, the Milky Way etc., indicated that one relatively small portion of the sky was more important than other portions even though specific constellations seemed not to be essential. Finally, Emlen (1972) found that the orientation was linked to

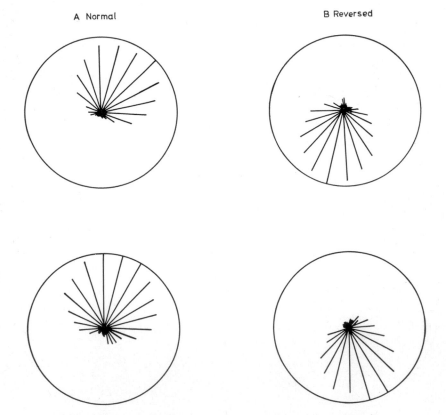

Fig. 28. Vector diagrams of indigo buntings; A, in spring under a spring planetarium sky; B, in spring under a spring planetarium sky horizontally rotated by 180°. From Emlen (1967a).

the rotation of the sky. The birds of one experimental group of buntings hand-raised without any sight of the sky were seemingly unable to orient in the ensuing autumn migration period. The birds of another experimental group were routinely exposed to a planetarium sky normally rotating around Polaris between fledging and the onset of autumn migration. These birds showed a southerly orientation, i.e. away from Polaris, during autumn migratory activity. The third group performed the crucial experiment: these birds were, between fledging and the onset of migration, routinely exposed to an experimental planetarium sky rotating around Betelgeuse, a star in the constellation Orion, which is in the southern sky in northern latitudes. The direction of the autumn migratory activity of these birds was northerly away from Orion, the pivot point of rotation. Thus, young indigo buntings, at least, between leaving the nest and autumn migration, watch the nocturnal sky to find out which part rotates least. In the northern hemisphere this is Polaris or the area around Polaris, which is then used as a reference for autumn migration. Exactly which part of the area around Polaris is still unsettled. For this star compass a clock is not required. It should be emphasized that this solution is in considerable contrast to that suggested by Sauer. A possibly undiscovered role of the earth magnetic field in Emlen's experiment is discussed in Chapter III, Section B, 5. Of course an equatorial or, even more so, a transequatorial migrant, could not rely on this compass for much of its trip. As discussed below (Chapter III, Section B, 5) there are suggestions how birds could cope with this problem. The majority of experiments dealing with star compass orientation are, however, based on the directionality of the spontaneous migratory restlessness. Training experiments also have a share in experiments on star compass orientation. Hamilton (1962a) trained ducklings of several species such as *Anas discors* and *Anas carolinensis* in a circular test apparatus (Fig. 29), during daytime, to look for water in a certain compass direction. When examined at night under the natural starry sky the ducklings were still able to select the training direction. They presumably used stars for this directional orientation. Wallraff (1969) used cardiac conditioning methods similar to those also used by Schlichte (Fig. 22) to train mallards (*Anas platyrhynchos*) under well-controlled planetarium skies. When the skies were familiar to the ducks through previous training sessions the birds reacted properly, but if they were exposed to a sky for the same location but some hours later, i.e. shifted in time, the birds were unable to recognize their training directions unless they were also trained during that part of the night or under that starry sky. Thus, there was no shift of direction according to shift of time as in sun compass orientation. Moreover, the mallards could be trained to entirely artificial star patterns which do not occur naturally. It may be con-

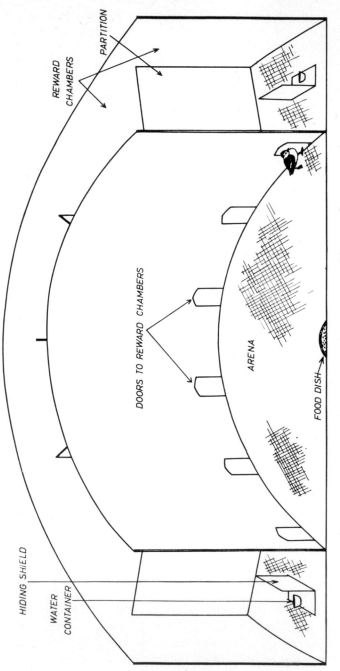

Fig. 29. Cross section of the experimental apparatus used by Hamilton (1962a) for directional training of ducklings.

cluded from these results that the star compass of mallards is learned, and requires considerable abilities of recognizing, distinguishing and memorizing complex star patterns. It does not involve apparent celestial movement nor chronometry to allow for movement. Unfortunately, the essential components of pattern recognition have not been analysed. I should like to emphasize at this point that all recent evidence indicates that stars are used for compass or directional orientation only. There is no confirmation whatsoever supporting the claims of Sauer and Sauer (1955, 1960) and Sauer (1957, 1961, 1963) for a star navigation system in European warblers (genus *Sylvia*) and golden plovers (*Pluvialis dominica*). For further discussion of this see Chapter VI, Section C.

4. The magnetic compass

At present the use of a sun and a star compass has been demonstrated for at least some birds. Another means of directional orientation has also

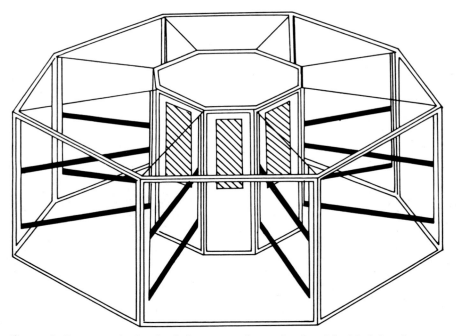

Fig. 30. A diagrammatic sketch of the octagonal cage developed by Merkel and Fromme (1958), used by these authors and later extensively by Wiltschko in experiments on magnetic compass orientation in European robins and warblers. After Merkel and Fromme (1958).

been documented, i.e. a magnetic compass. Electro-mechanical recording methods were used. Merkel and Fromme (1958) introduced an octagonal cage with eight radial perches (Fig. 30). The experimental bird—mostly European robins (*Erithacus rubecula*) in the laboratory of Merkel and later of Wiltschko—can move freely around the central structure. Each hop is recorded by one of the eight radial perches. With robins this arrangement proved better than the circular perch system (that is in fact an automatic-ally-recording extension of Kramer's classical circular cage) as used by, e.g. Hamilton (1962b), Perdeck (1963), Mewaldt *et al.* (1964) and Wallraff (1966a, 1972a) in experiments on sun and star compass orientation. The scores recorded by the radial perches in one night are processed vectorially for mean direction. The means of several nights and/or several birds are summarized and subjected to (second order) statistical analysis.

The directional ability of migratorily restless European robins without visual cues was first established by Merkel and Fromme (1958) and Fromme (1961) and later confirmed by Merkel and Wiltschko (1965), and Wallraff (1972a), provided the birds were allowed a period of adaptation. The directions recorded agreed largely with the ancestral migratory direction. Merkel and Wiltschko (1965) suggested that the earth magnetic field was used for orientation. Figure 31A presents a summary of 97 "bird nights" in the natural magnetic field in spring. When an artificial magnetic field was applied, magnetic north rotated to geographical east-south-east (Fig. 31B) or west (Fig. 31C), and the bird's direction followed accordingly (Wiltschko 1968, 1973). Figure 32 shows the arrangement of Helmholtz coils around an octagonal test cage as used extensively by Wiltschko.

Magnetic orientation in birds (and the evidence of Wiltschko's experi-ments) has long been doubted and questioned for the following reasons.

(1) Orientation in one "bird night" is usually weak, i.e. the Rayleigh test, if applied despite the lack of independence among the data, often indicates randomness and rarely reaches the 1% level. Significance is usually attained by second-order statistics.

(2) Repetitions, at least for a while (e.g. Perdeck, 1963; Emlen, 1970) seemed as unsuccessful as initial attempts to demonstrate magnetic effects on homing pigeons—they all turned out later to have failed for methodological reasons.

(3) Nobody had been able to obtain conditioned responses of birds to magnetic fields.

(4) It was entirely unknown how the earth's weak magnetic field could be perceived by birds, as an obvious magnetic organ does not exist.

Some details of how the earth's magnetic field may be used by robins were found by Wiltschko and Wiltschko (1972) it is possible that the bird's magnetic compass does not use the polarity of the field, as does a

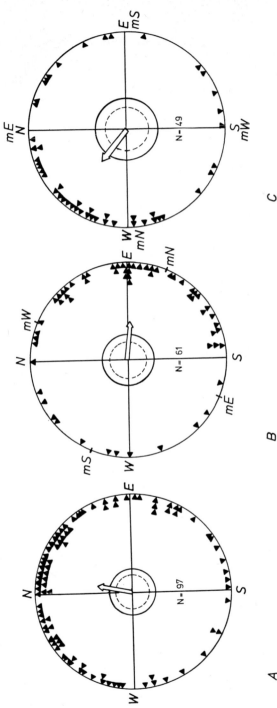

Fig. 31. Spring orientation of European robins (A) in the normal earth magnetic field, (B) upon experimental rotation of magnetic North (mN) to east-south-east, or (C) to just south of west. Each black triangle represents the mean direction of one bird night; N indicates the total number. The centrifugal arrow represents the mean of each summary. The inner concentric circle (dashed) represents the 5%, the second circle (solid) the 1% critical vector length (Rayleigh Test) for uniformity. From Wiltschko (1973).

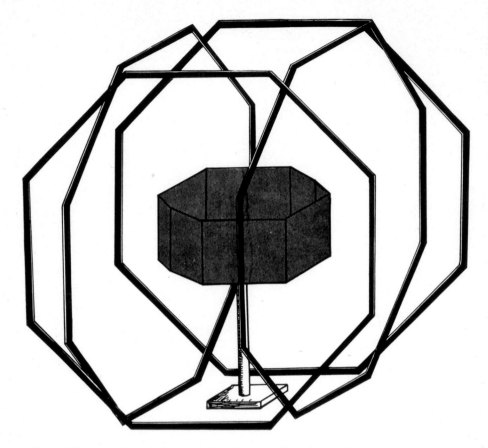

Fig. 32. The arrangement of two Helmholtz coils (diameter = 180 cm) around a test cage, as used extensively by Wiltschko (1968).

man-made magnetic compass—it seems to evaluate the inclination of the axial direction of the field lines to the vertical. More precisely, the birds take the smaller angle between field lines and the vertical as "pole-ward". This surprising result was obtained in the following way: a group of robins was tested under five different magnetic conditions during spring migratory activity. As a control experiment (Fig. 33A), the unchanged local field of Frankfurt was used and the birds were oriented in northerly directions. In a series of experiments shown in Fig. 33B the polarity of the field was reversed: the birds continued to be oriented in northerly directions. In the experiments shown in Fig. 33C the vertical component and in Fig. 33D the horizontal component was reversed, both resulting in the smaller angle between field lines and the vertical pointing

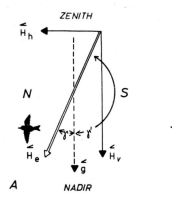
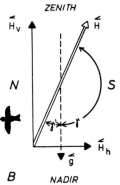

A   NADIR   B   NADIR

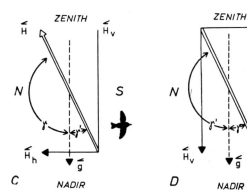

C   NADIR   D   NADIR

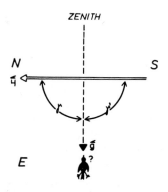

Fig. 33. Diagrammatic representation of experiments on the inclination compass by Wiltschko and Wiltschko (1972).

southward: the birds were oriented in southerly directions. In experiments shown in Fig. 33E the magnetic field was horizontal: there were only two right angles and the birds were unable to orient themselves. This inclination compass has one advantage over a polarity compass: the polarity of the earth's magnetic field has reversed repeatedly in the past and a polarity compass would have grossly misled the birds.

For an inclination compass, however, reversal of polarity means no change at all (as shown in Fig. 33A and B), demonstrating its evolutionary advantage over a polarity compass. On the other hand, an inclination compass is useless at the equator where field lines run horizontally (Fig. 33E), and it also reads towards the pole in the southern hemisphere (Fig. 33D) (i.e. southward). Thus, an inclination compass is fine if the birds stay in one hemisphere. The European robin is such a bird, but there are numerous other birds that migrate to equatorial regions or even beyond the equator. They may well cope with the problems involved by temporarily turning to other, for example celestial means, for compass orientation. This interrelation or hierarchy of compasses is discussed in more detail in the next chapter.

One of the long-standing arguments against a magnetic compass was that the energy involved was too low to be utilized by a bird. This idea of the sensitivity of avian magnetic reception was wrong. In fact, the European robin's magnetic compass turned out to be adapted to an amazingly narrow range of intensity. Wiltschko (1968, 1972, 1978) demonstrated experimentally that his birds could orient in an intensity range from 0·40 Oe to about 0·57 Oe. They were temporarily disoriented when total field intensity was decreased from the natural 0·46 Oe, or increased above the limits of 0·40 Oe and 0·57 Oe respectively. After a period of three days adaptation to the new intensity the birds were again oriented properly. Wiltschko (1978) also succeeded in adapting robins to two different intensities of 0·16 Oe and 1·5 Oe at the same time. However, the birds turned out not to be adapted to the intermediate range between 0·6 Oe and 1·5 Oe. This adaptability is an essential property of the system since the total intensity of the earth's magnetic field also changes latitudinally within the range of bird migration.

These findings are essential details of the mechanism of magneto-reception but the sensory mechanism itself still remains obscure. As birds react to "magnetic storms", which are in fact small variations of the geomagnetic field, e.g. from 0·460 Oe to 0·465 Oe (Southern, 1969, 1978; Keeton *et al.*, 1974), the avian magnetic compass appears to be sensitive to changes of at least 0·005 Oe. Theoretical possibilities of magneto-reception, such as utilizing the Hall effect in several different ways, have been entertained but never documented. New impulses in the search for the physiological mechanism of magneto-reception were provided by Leask (1977, 1978). According to this model, magnetic fields may be perceived as a by-product of the normal visual process through rhodopsin molecules or molecules similar to rhodopsin. The energy required would be provided by optical pumping. Thus, this part of the model can easily be tested by finding out whether or not birds respond to magnetic fields in total darkness. In reviewing previous experiments to the extent possible from the literature or personal communication, there was, in all the experiments, some residual light, or, in Wiltschko's experiments, the birds were introduced to the test cage before dusk. Instances in which orientation was absent or at least questionable such as in some of Emlen's repetitions (Emlen, 1970; see also Emlen, 1975, pp. 199–203) birds were placed into the test cages after dusk. Anyway, the light requirement of Leask's model can be easily and quickly tested. The necessary energy may also, however, be provided by biochemical processes and a negative result in this respect may not discount the model.

One of the reasons for scepticism about the magnetic compass theory was the failure of attempts to confirm sensitivity of birds to magnetic fields in laboratory conditioning experiments. After many abortive attempts (e.g. Griffin, 1940, 1952; Kramer, 1949, 1950a; Clark *et al.*, 1948; Orgel and Smith, 1954; Fromme, 1961; Meyer and Lambe, 1966; Emlen, 1970; Kreithen and Keeton, 1974c; Beaugrand, 1976), the only exception being Reille (1968), (whose results were possibly based on incorrectly calculated field strength as discussed by Kreithen and Keeton, 1974c), Bookman (1978) finally reported success. He trained pigeons to discriminate between the presence or the absence of a 0.5 Oe vertical field, introduced by Helmholtz coils, in a flight tunnel inside a mu-metal shielded chamber. The birds rarely flew through the tunnel; they usually walked. Successful discrimination was, however, associated with spontaneous flutter activity such as hovering, jumping or turning, which may be the reason why previous attempts, using strapped-down or restricted pigeons or birds, failed.

5. Interrelation or hierarchy of compasses

The sun compass was, historically, the first compass discovered in birds, and it is also perhaps the most prevalent. If the sun is available, birds rely on the sun compass. Sun compass orientation is, in contrast to magnetic compass orientation, easy to demonstrate in fairly simple experiments. Hoffmann (1953) published one attempt to demonstrate that the essential components of the sun compass were innate. A starling taken from the nest box at an early age and hand-reared indoors, later performed properly in food-rewarded sun compass training. The experiment has never been repeated. Recently evidence has begun to show that, at least in homing pigeons, the sun compass is learned and possibly calibrated relative to the magnetic compass (Wiltschko *et al.*, in preparation).

In contrast to the orientation of experienced birds in the sun, the orientation of inexperienced young pigeons was affected by magnets (Keeton, 1971a). Moreover, pigeons raised without sight of the sun were able to orient properly when released under overcast (Wiltschko *et al.*, 1976). These findings suggested that the magnetic rather than the sun compass may be the primary compass. Recent experimental results indicate that the sun compass is learned rather than innate as previously assumed. Wiltschko *et al.*, (1976) reared pigeons in an artificial photoperiod which was constantly shifted 6 h counter-clockwise. The birds were given free flight and short-distance training flights during the afternoon, i.e. the common light period of the artificial and the natural day. When released, experimental birds and controls were oriented towards home. After resynchronization of the experimental photoperiod to natural conditions, the experimental birds performed like birds after a shift 6 h clockwise and deviated counter-clockwise at vanishing. In later releases the sun compass turned out to be readjusted to natural conditions and the birds could now orient correctly. Hence, the birds' association of time, sun azimuth and geographic direction, appears to be established in a learning process. It may be readjusted and it does not seem to be innate. Experiments are still in progress but it may be assumed with some certainty that sun compass is calibrated on the basis of the magnetic compass.

Returning to migratory birds, evidence has been accumulated with European robins and European warblers, in a series of experiments by Wiltschko, that the basic compass may be the magnetic compass and the star compass the secondary system calibrated relative to the magnetic compass. In a series of experiments (Wiltschko and Wiltschko 1975a,b; 1976a) in Spain, the long-distance migrating garden warbler (*Sylvia borin*) and the whitethroat (*Silvia communis*) and the short-distance migrating subalpine warbler (*Sylvia cantillans*) were first tested under the natural starry sky in the natural magnetic field. Figure 34A–C presents the results for the garden warbler. The directions indicated in Wiltschko cages agreed with the ancestral direction of migration. Upon rotation of the magnetic field by 120° clockwise, magnetic north was now pointing east-south-east and all three species shifted their direction according to the reset magnetic field and ignored the stars. If the horizontal component of the magnetic field was compensated, depriving the warblers' inclination compass of cues for orientation, the overall performance was random, although stars were available. If the data of Fig. 34C were, however, subdivided into two groups acccording to whether the birds had previously served as control or experimental birds (black and white symbols in Fig. 34C), some tendency

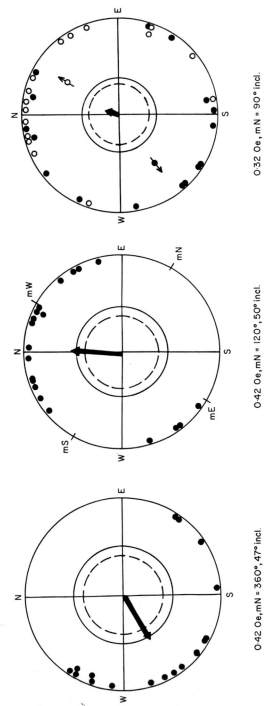

0.42 Oe, mN = 360°, 47° incl.

A

0.42 Oe, mN = 120°, 50° incl.

B

0.32 Oe, mN = 90° incl.

C

Fig. 34. Directional responses of autumn migratory garden warblers (*Sylvia borin*) under the natural starry sky: A, in the natural magnetic field; B, in a magnetic field rotated 120° clockwise and; C, with horizontal component compensated. Each symbol indicated the mean direction of one bird in one night. Subsamples of C are explained in the text. From Wiltschko and Wiltschko (1975a).

emerged suggesting that the direction chosen by the birds with respect to the stars was related to the magnetic conditions of the previous experiment. This would possibly indicate the use of the stars as a secondary directional reference recalibrated from the magnetic field.

European robins were also subjected, during spring migration at the same site in Spain, to the same experimental procedure. The results were somewhat different: upon rotation of the magnetic field 120° clockwise the robins at first continued in their northerly direction for two nights, then shifted about 120° clockwise. In the partially compensated magnetic field, which was useless for orientation by an inclination compass, birds previously used as controls maintained northerly directions, while those previously used as experimental birds first preferred northerly directions and then shifted to the east-south-east. The concept of stars being calibrated or recalibrated on the basis of magnetic information was further confirmed in laboratory experiments during spring migration using an arbitrary pattern of 16 "stars" (Fig. 35). Figure 36 documents clear-cut results. The reduced magnetic field and the artificial "star" pattern, which were both new to the birds, caused disoriented behaviour (Fig. 36A). In a magnetic field of normal intensity rotated 80° clockwise, the robins were properly

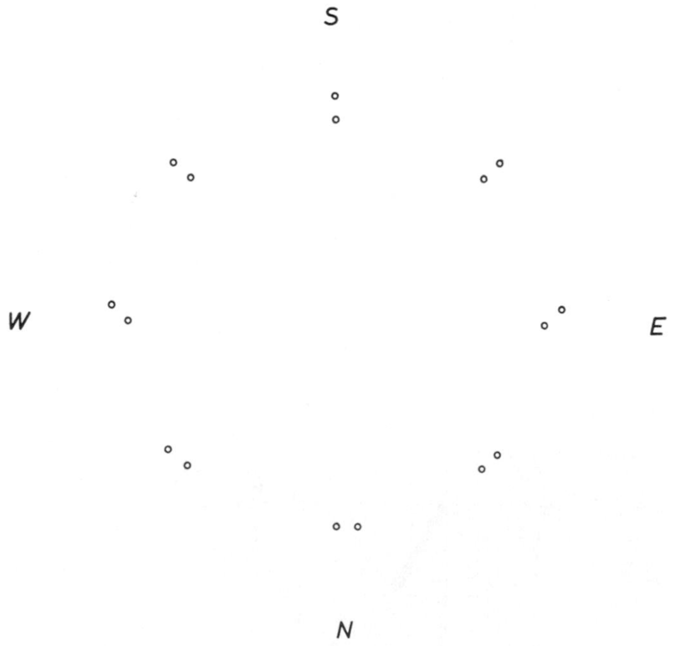

Fig. 35. The artificial pattern of 16 "stars" used by Wiltschko and Wiltschko (1976a) in laboratory experiments on the calibration of the star compass.

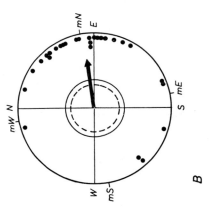

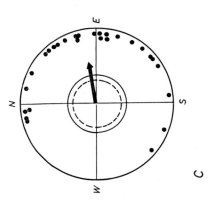

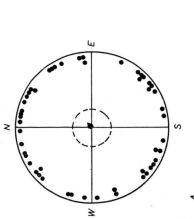

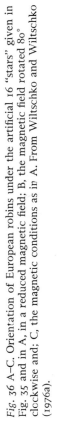

Fig. 36 A–C. Orientation of European robins under the artificial 16 "stars" given in Fig. 35 and in A, in a reduced magnetic field; B, the magnetic field rotated 80° clockwise and; C, the magnetic conditions as in A. From Wiltschko and Wiltschko (1976a).

oriented to magnetic North (Fig. 36B). During this period they apparently calibrated their star compass since they continued to be oriented in the same direction (Fig. 36C) after the magnetic field had been returned to the conditions of the series given in Fig. 36A, that had been unable to produce oriented behaviour.

The differential response of warblers and robins may have migratory significance and be of adaptive value. Warblers, as long-distance migrants, experience drastic changes of the nocturnal sky between breeding and winter ranges. In addition transequatorial migrants have to pass a zone where the lines of the geomagnetic field run horizontally (Fig. 37). The inclination compass is therefore, temporarily unusable. It would be advantageous for them to be able to check and to recalibrate their star

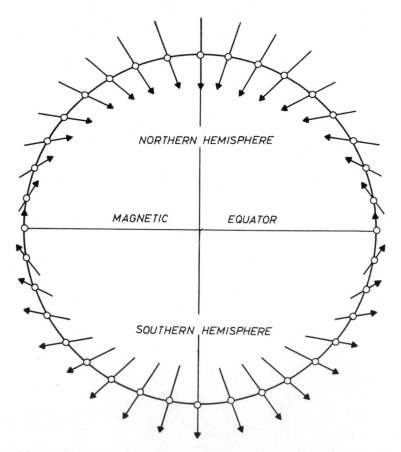

Fig. 37. Diagrammatic view of the direction of the geomagnetic field at the surface of the earth in different latitudes. Different length of arrows indicate different intensities. From Wiltschko and Wiltschko (1976b).

compass repeatedly. For European robins, as short-distance migrants, the starry sky does not change much during migration and they do not need to check and recalibrate their star compass.

A star compass is evidently an advantage: it is probably easier to maintain a heading with the help of stars. Even in cage experiments orientation under stars is usually superior (in terms of less scatter) to solely geomagnetic orientation, which is mostly so weak that only second-order statistics reveal significant directedness.

The establishment of the star compass as found by Wiltschko is different from the process described by Emlen. These differential findings do not necessarily contradict each other. At the time of Emlen's experiments, magnetic conditions were not considered and it is possible that the magnetic field influenced Emlen's results without having been discovered. Further careful experiments will reveal whether some birds such as the indigo bunting use different systems to establish a star compass, compared to birds such as European robins and warblers.

Astronomical compasses have the disadvantage of operating only under clear or partly clear skies. With the exception of a rather narrow band at the magnetic equator the magnetic compass is available independent of meteorological conditions and time of day. Magnetic "storms", in fact variations in the third decimal place, e.g. from 0.460 Oe to 0.465 Oe, have been shown to interfere with the orientation of gull chicks (Southern, 1969, 1978) and homing pigeons (Keeton *et al.*, 1974). They are, however, rare and effects on migrating birds are only occasionally discovered (Richardson, 1974; Moore, 1977).

6. Compass and distance

Compasses indicate directions but not goals. If a bird "looks at" one or two or even three of its compasses it can select and maintain directions, alter its heading and so on, but it cannot identify the direction of a goal. If the bird has some sort of a map as outlined by Kramer (1953) showing the relationship between its present location and its goal, the bird could align the map with a compass, read the direction and possibly aim for home and take off. Information on direction and possibly distance may also be provided in some other way, i.e. genetically.

There is good evidence, as discussed in Chapter IV, Section B and Chapter VI, Section G, that at least in some species of migratory birds, the young reach their winter range in the autumn on the basis of two sets of genetical information, one on direction, the other on distance. Compasses which are entirely sufficient to explain all directional performances have been discussed in the preceding paragraphs. The distance is a more

complex problem, having been only recently tackled in laboratory experiments with old-world warblers of the genera *Phylloscopus* and *Sylvia*
by Gwinner (1968a,b, 1969, 1972, 1974) and by Berthold *et al.* (1972) and
Berthold (1973, 1978).

In many series of experiments with migratorily restless warblers, the
authors demonstrated that migratory restlessness and related phenomena
such as, e.g. moult, are functions of endogenously controlled circannual
rhythms. The authors compared the amount of migratory restlessness
expended in the recording cages among species with differentially long
migration routes. A relationship was found between migratory restlessness
measured as time or as amount of activity, and migratory distance of
free-living conspecifics: they were proportional. Taking three warblers
of the genus *Phylloscopus* as an example, the long-distance migrating
willow warblers (*Phylloscopus trochilus*), wintering in southern Africa
(Fig. 38), spent proportionally more time or activity than wood warblers
(*Phylloscopus sibilatrix*), wintering in central Africa, and those in turn
spent proportionally more than chiffchaffs (*Phylloscopus colybita*) wintering in northern and northeastern parts of Africa. In fact, quantitative
aspects of these experiments were sufficiently precise to permit Gwinner
(1972) to calculate the theoretical distance equivalent to the total
amount of restlessness produced in the laboratory. The calculated distances
agree surprisingly well with those actually covered in migration.

These results support the view that an endogenous temporal programme
determines the duration (or amount) of migratory activity, and this is
equivalent to information on distance. How this mechanism works physio-
logically, how the birds measure activity, energy expenditure or distance
is entirely unknown. The accuracy of such an indirect system certainly
cannot be expected to be overwhelming. It is, however, well documented
by banding recoveries that young birds are less accurate than experienced
older individuals. In principle, a system as outlined above, using information on the direction plus information on the distance to travel, would
work and would explain one aspect of bird migration: how a young bird
reaches the winter range during its first autumn migratory season. Field
experiments supporting these views are discussed in Chapter IV. Section B.

It is well known that migratory pathways of many birds are not just
straight lines but that one or perhaps more changes of direction occur
en route. Routes around the Mediterranean from central Europe east by
way of the Bosporus or west by way of Gibraltar may serve as examples.
Gwinner and Wiltschko (1978) tested the direction indicated by garden
warblers (*Sylvia borin*) during their autumn migratory season, who were
exposed to the natural magnetic field in Merkel–Wiltschko cages under
constant light (12 : 12 LD) and temperature (20° ±2°C) conditions. As
may be seen from the data presented in Fig. 39 the warblers headed

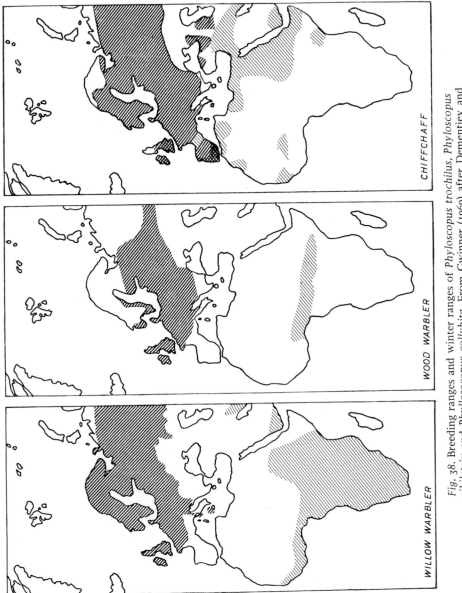

WILLOW WARBLER   WOOD WARBLER   CHIFFCHAFF

Fig. 38. Breeding ranges and winter ranges of *Phylloscopus trochilus, Phylloscopus sibilatrix* and *Phyllosocopus collybita*. From Gwinner (1969) after Dementiev and Gladkow (1954).

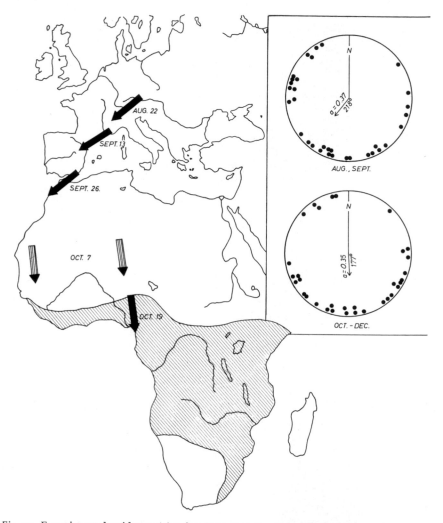

Fig. 39. Experimental evidence (circular diagrams) for an endogenous control of changes of the migratory direction in garden warblers (*Sylvia borin*). In the circular diagrams each score indicates the mean diriection of one bird in one night of those birds that scored in the August and September as well as in October–December testing period. The arrows on the map indicate directional preferences at those dates at which free-living garden warblers pass through banding stations along their migratory route from central Europe to the winter range in Africa (shaded area). From Gwinner and Wiltschko (1978).

to the south-west in August and September and shifted their direction to just east of south in October, November and December, in agreement with the timetable of natural outdoor migration. These results supported the view that endogenous timing-processes control migration and may even account, at least in part, for curved migratory routes. Undoubtedly, such endogenous mechanisms should not be expected to be very accurate. Conceivably, as pointed out by Gwinner and Wiltschko (1978) the endogenous mechanism may well be just one integral component of a complex system not yet known in more detail. It should be mentioned here that Sauer (1957) reported observations of a corresponding shift of a lesser whitethroat (*Sylvia curruca*) in planetarium experiments in which the planetarium sky was, successively, adjusted to southern latitudes. This species flies round the Mediterranean via the eastern route. As mentioned elsewhere, (Chapter VI, Section C) Sauer's data are not convincing because of extremely small sample sizes and for methodological reasons precluding statistical treatment. Taken at face value, however, Gwinner and Wiltschko (1978) argue that the more southerly headings were also observed later in autumn. Hence, the southward shift—methodological considerations apart—that was attributed by Sauer to star navigation, may support the proposed concept of endogenous timing-control and the use of geomagnetic cues.

IV. Experimental field work with wild birds

The work to be reported and discussed in this first section consists of displacement experiments of various types. In some, birds have been displaced from their nests in homing experiments very similar to those with homing pigeons (Chapter V) and in other experiments birds have been caught and displaced during migration. Some deal with displacement experiments of non-migratory birds. Only very few experiments involve direct experimental interference with the proposed mechanism of orientation.

A. Displacements from breeding or wintering ranges

Adult birds taken from their nests in homing experiments revealed pin-point navigational abilities as confirmation—if any is needed—of the same qualifications demonstrated season after season by migrants returning to the same nest after extensive migration. Before listing and discussing some examples I would like to emphasize the technical problems encountered in homing experiments with free-living birds.

(1) Experiments are restricted to the breeding season.
(2) It is frequently very difficult to identify the experimental bird and to time its arrival and sometimes it remains uncertain whether or not the bird homed at all (it could have returned to a nearby location not under observation).
(3) It is often impossible to assess the motivational state of the bird, i.e. whether or not the experimenter's assumption is correct that his bird wants to home.

(4) Many wild birds are difficult to catch and keep in captivity to carry out displacement or additional experimental manipulations, such as clock-shifts, surgical interactions etc., without critically affecting their physical condition or motivation.

One of the most spectacular examples of successful homing is that of a Manx shearwater (*Puffinus puffinus*) which was displaced from its burrow on Skockholm (off the south-western corner of Wales) to Boston, Massachusetts, and which returned home in only 12½ days, over more than 4500 km (Matthews, 1953a). The reverse experiment was carried out with seven Leach's petrels (*Oceanodorma leucorhoa*) taken from their homes on Kent Island, New Brunswick, Canada and released from the coast of Sussex, England, 4800 km away. Four homed, two of them within less than 14 days (Billings, 1968). Kenyon and Rice (1958) displaced Laysan albatrosses (*Diomedea immutabilis*) from Midway Island as far as 6600 km to Luzon. One albatross negotiated the distance back to Midway in 32 days. These three species naturally roam widely while foraging and the experimental success may therefore not be as impressive as it first appears. But homing experiments have also been performed on numerous species with restricted foraging ranges during the breeding season. To select a few recent examples from, and to extend, Matthews's list (1968): Southern (1959, 1968) displaced purple martins (*Progne subis*) and Downhower and Windsor (1971) bank swallows (*Riparia riparia*), following a long history of similar experiments associated with the names of pioneers like Rüppell, Griffin, Watson, Lashley and others.

In general, percentages of homing have been low and homing speeds slow, but this does not necessarily mean that the birds used were poor navigators. It may simply reflect the methodological problems incurred in these experiments.

Birds have also been displaced from their winter quarters. If displacement was within reasonable distances some homing within the same winter season was recorded by several authors. According to Roadcap (1962), of 81 white-crowned sparrows (*Zonotrichia leucophrys*) and golden-crowned sparrows (*Zonotrichia atricapilla*) wintering in California, 17 returned from displacements 15–260 km away, some across mountain ranges. Ralph and Mewaldt (1976) displaced 905 sparrows of the same two species from their Californian winter homes 5–160 km away. Ninety-seven returned (judging from the time involved not by the most direct route), while 137 remained near the release sites. In accordance with the results of Ralph and Mewaldt (1975), a higher proportion of sub-adults stayed, and a higher proportion of adults returned. The implications of these findings are discussed in Chapter IV, Section C. Birner *et al.* (1968) displaced starlings, but these were mostly not wintering migrants, but banded non-migratory members of the local breeding population.

Mewaldt (1963, 1964a,b) shipped more than 1000 wintering sparrows of the genus *Zonotrichia* (*Z. leucophrys gambelii*, *Z. leucophrys pugetensis*, and *Z. atricapilla*) from their established wintering grounds at San Jose, California to either Baton Rouge, Louisiana (2900 km east-south-east), or to Laurel, Maryland (3860 km due east) (Fig. 40) and some returnees

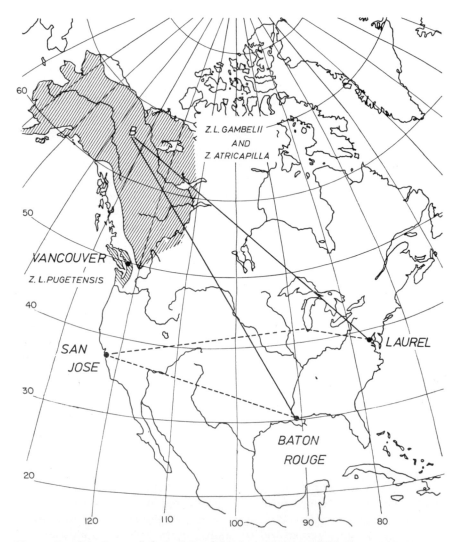

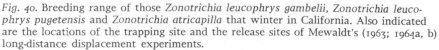

Fig. 40. Breeding range of those *Zonotrichia leucophrys gambelii*, *Zonotrichia leucophrys pugetensis* and *Zonotrichia atricapilla* that winter in California. Also indicated are the locations of the trapping site and the release sites of Mewaldt's (1963; 1964a, b) long-distance displacement experiments.

from Baton Rouge (to San Jose next winter) to Laurel the next time. No displaced bird returned to the trapping site at San Jose during the same season but significant numbers returned for the next and subsequent wintering seasons. For example, from 411 birds displaced to Baton Rouge in 1961–62, 26 were recaptured in 1962–63, and from 660 displaced to Laurel in 1962–63, 15 were recaptured in 1963–64. Of 22 birds displaced to Laurel after they had returned from Baton Rouge, six were retrapped again at San Jose. Probably all returnees returned by way of spending the breeding season in their breeding range in Alaska and not by direct homing. Nevertheless, this feat is remarkable in terms of physical performance as well as in terms of navigational performance. It is interesting to note in this context that some of the *Zonotrichia leucophrys pugetensis* breed only as far north as Vancouver, B.C. and, perhaps because they are naturally less extensive migrants, some returned from Baton Rouge but none from Laurel.

If *Zonotrichias* were displaced to exceedingly distant locations such as Seoul, Korea (9060 km great-circle distance westward), none ever returned (Mewaldt *et al.*, 1973). The two species involved, *Zonotrichia leucophrys gambelii* and *Zonotrichia atricapilla*, undertake over-water flights during normal migration and the authors hypothesized that these strong migrants would depart from Korea towards the north-east, fly along the east coast of Asia, possibly cross the Bering Sea to reach their North American breeding grounds and eventually go to San Jose.

B. Displacements during migration

As already discussed in the preceding section, bird banding may become particularly rewarding and informative if birds are not just banded and released but if banded and released upon some displacement. This holds especially true for Perdeck's work on starlings, clearly showing what was already suggested and outlined by earlier results associated with the names of Drost, Rowan, Rüppell, Schüz, Bellrose and others.

Modifying and extending the previous work of van Dobben (1939) and Klomp (1949, 1950), Perdeck (1958), between 1948 and 1951, caught more than 11 000 starlings (*Sturnus vulgaris*) during their autumn migration in Holland. Previous banding results had revealed that the migrants' breeding range extended around the Baltic Sea as indicated in Fig. 41. They usually winter in southern England, southern Ireland and northern France as shown in Fig. 41. After determining their age and banding them, Perdeck displaced the starlings to Switzerland by aeroplane and

Fig. 41. Recoveries of breeding and of wintering starlings (*Sturnus vulgaris*) banded during autumn migration in The Hague, Holland. From Perdeck (1958).

released them there. More than 300 recoveries coming in over the years revealed the navigationally differential behaviour of adults and juveniles. Juveniles having been displaced during their first autumn migration continued to maintain their ancestral direction to the west-south-west, essentially parallel to the direction seen in Holland (Fig. 42). Recoveries in subsequent months and years indicate that the juvenile starlings continued from Switzerland roughly as long or as far as they would have had to travel from Holland. By ending up in Spain Perdeck's displacement had established a new winter range for the Baltic population, though displaced juveniles continued to winter in Spain. In subsequent displacements of starlings from Holland directly to Barcelona, carried out by Perdeck (1964, 1967), juveniles continued to migrate towards the west-south-west (Fig. 43), in agreement with the findings from the displacement to Switzerland. The majority of adults may, as judged by a number of short distance recoveries, initially have kept a west-south-west course, then headed toward the ancestral winter range.

Thus Perdeck's experiments document particularly clearly what earlier work by others with other bird species such as storks, crows and hawks had outlined: juveniles of many migratory species reach their winter range for the first time by possessing, probably genetically, information

Fig. 42. Recoveries of adult and of juvenile starlings displaced to and released from three airports in Switzerland during autumn migration. From Perdeck (1958).

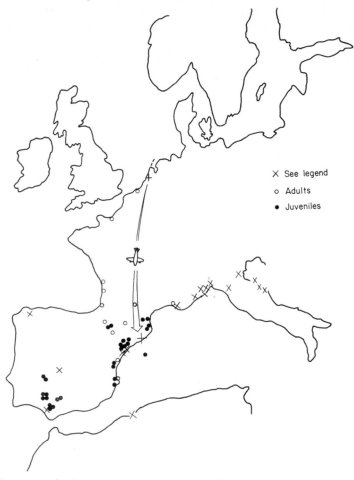

Fig. 43. Autumn and winter recoveries (circular symbols) of adult and juvenile starlings after displacement from Holland to Barcelona, Spain. Crosses indicate recoveries of juveniles during spring migration following autumn displacement. From Perdeck (1967).

on direction and distance. Thus, the results from field work support the results from laboratory work by Gwinner and by Berthold reviewed in Chapter III, Section B, 6, and vice versa.

The question remains of where the experimental juvenile starlings settled for breeding, and this includes the question of the migratory direction taken in spring and also in the autumn of subsequent years back to the winter range. The juvenile starlings displaced to Switzerland and wintering in Spain, i.e. considerably to the south-west of the ancestral winter range, returned to the ancestral breeding area without any significant shift to the south or west. This is in agreement with findings discussed in Chapter IV, Section C indicating that the home range may be established in juvenile birds in some form that may be called navigational imprinting, in the period between fledging and departure for autumn migration. It may be extrapolated from suitable recoveries that Perdeck's experimental juveniles did not follow a straight migratory route but rather a route with a change of heading. An answer to this question is important for the phylogenetic area of bird migration, since many experimental resuits (including those of Perdeck), indicate that the initial migratory direction is genetically fixed.

The relevant recoveries of birds from Perdeck's displacement experiment to Barcelona accumulated in an east-north-easterly direction. mostly in northern Italy (Fig. 43). If extended, i.e. if the starlings continued in that direction, one would have to assume that the starlings were heading for a breeding area considerably to the south of where they were hatched, which extrapolating from other experimental results, had presumably been imprinted on them. However, only three recoveries were made during the breeding season, one each from Yugoslavia, south-eastern France and Denmark, which is not sufficient to reach a definite conclusion.

Extensive as Perdeck's experiments were, despite their high yield of important contributions to various aspects of bird migration and orientation, essential questions still remain.

C. Establishment of breeding and of wintering ranges

Perdeck's experiments with starlings illuminated the migrational aspects of how wintering ranges are established. They also touched on the question of how breeding ranges are established. Some experiments have dealt directly with the question of breeding ranges, e.g. Mauersberger (1957) reported on attempts made in Russia to stock with pied flycatchers (*Ficedula hypoleuca*), areas that were previously not inhabited by this species.

Shcherbakov and Polivanov (in Mauersberger, 1957) hand-reared almost 1400 flycatchers and displaced them 50 and 400 km to the south to suitable habitats before, or shortly after, fledging. About 6·5% of the displaced young flycatchers returned in later years, successfully establishing a breeding population at the new location.

Löhrl (1959), working in south-western Germany, checked the extent of the sensitive period in which "home" seemed to be established, in more detail. He displaced a total of 134 juvenile collared flycatchers (*Ficedula albicollis*), in two consecutive years, 90 km to the south about 2–3 weeks before the onset of migration. This onset cannot, however, be precisely determined. In subsequent years 19% of the displaced flycatchers returned from migration to the immediate vicinity of the experimental release site, also establishing a new population at a location previously uninhabited by this species. In another series of experiments 68 juveniles were displaced and released when the onset of migration was imminent. No bird was ever recovered, but local circumstances were less favourable than in the first series of experiments and may, at least in part, account for the failure.

Berndt and Winkel (1978) exchanged more than 300 eggs, nestlings and juveniles of the pied flycatcher (*Ficedula hypoleuca*), between two suitable study areas 250 km apart in northern Germany. Without exception experimental birds later settled for breeding where they had fledged or had been released at the age of 36 days, after having been hand-reared in the other area. These results are entirely in agreement with those of Shcherbakov and Polivanov and or Löhrl.

A very short period of time (possibly as little as 10 days of exposure) suffices to establish "home" in navigational terms ("navigational imprinting"), i.e. the location to which the bird will return from its first migration next spring and in subsequent years. A similar sensitive period was described by Emlen for establishing the star compass in indigo buntings (Chapter III, Section B, 3). A corresponding period is known to pigeon breeders: if one wants to transplant young pigeons to some new location it can be done most easily immediately after fledging ("weaning") and will be increasingly unsuccessful if done later and finally becomes almost impossible.

A similar process of "navigational imprinting" seems to take place for the winter range during the early phases of the first wintering season of migrants. Of over 900 wintering white-crowned (*Zonotrichia leucophrys*) and golden-crowned (*Zonotrichia atricapilla*) sparrows displaced by Ralph and Mewaldt (1975) in California 4–160km away, no birds displaced as adults returned to the release site next winter, while many returned to the original capture station. Of the birds displaced as sub-adults, however, some returned before mid-January to the vicinity of the new site in the following year, while none of those displaced after mid-January were recaptured at the new site.

D. Initial orientation of displaced birds

In the preceding chapters evidence was largely based on recoveries of tagged birds. There is also extensive work on the initial orientation taken up by displaced birds when released. Bellrose (1958) trapped mallards (*Anas platyrhynchos*) as passage migrants and presumably also as wintering individuals, at several locations in Illinois, USA, and released them from various sites 18–53 km away. Under overcast skies, day or night, random initial orientations were observed. Under clear skies throughout the year day and night, initial orientation was directed toward the north-north-west. Those being on autumn migration were found to continue their southward migration subsequent to Bellrose's experiment. The reason for this strange initial north-west tendency, seemingly without biological significance, is as little understood as the similar initial north-west tendency (or lack of orientation under overcast) found by Matthews (1961) in non-migratory mallards in south-western England. Matthews coined the term "nonsense orientation" for this phenomenon. In his first major series of experiments Matthews (1961) accumulated a year-round total of 714 vanishing points for mallards from Slimbridge, England. Figure 44 shows the resulting impressive circular distribution with its mean vector at 314°. Subsequent investigations which accumulated several thousand scores (Matthews, 1963a), revealed that the apparent similarity between Bellrose's and Matthews' findings was not

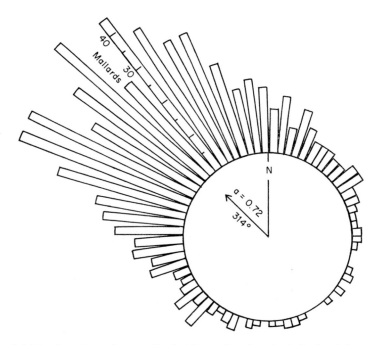

Fig. 44. Initial orientation of 714 mallards (*Anas platyrhynchos*) displaced from Slimbridge, England, into various directions and distances. From Matthews (1961).

due to a general or a species-specific phenomenon in mallards. Mallards from London took off in southerly directions and the initial orientation of mallards from Peakirk differed from that of Slimbridge mallards. Summer and early autumn orientation was roughly like that of the Slimbridge birds (perhaps slightly more westerly), and late autumn, winter and spring orientation was nearly random. The latter may be partly explained by an influx of migratory mallards in that season. Swedish mallards showed a south-easterly tendency at Stockholm. Though attempts to rear mallards from various origins in aviaries and to test their "nonsense orientation" failed, there are indications that specific "nonsense" tendencies may have a genetic basis. Matthews (1967) also showed that this stereotyped initial orientation is a short-lived affair: the birds abandon their initial directionality within a few kilometres or minutes, and its significance remains as obscure as at the outset. It is a type of orientation behaviour without any obvious sense.

"Nonsense" orientation deserves the consideration given to it here because it is an orientational phenomenon, and also because Matthews used it in his attempts to reinterpret the results of others in pigeon homing experiments which were not in agreement with his hypothesis on sun navigation (to be discussed in Chapter VI, Section B.)

E. Direct experimental interactions

Due to the methodological difficulties mentioned earlier, wild birds have only rarely been used in direct experimental interactions such as clock-shifts or surgical interference in displacement experiments, while many such experiments have been carried out with pigeons.

Bellrose's and Matthews' evidence of disorientation under overcast has already suggested that their ducks used the sun compass during daytime and the star compass at night. Matthews (1963b) confirmed the utilization of the sun compass by mallards in clock-shift experiments (Chapter III, Section B, 2 and Chapter V, Section A); there were roughly 90° counter-clockwise deviations of initial orientation with a clock-shift of 6 h clockwise; 90° clockwise deviations of initial orientation with a clock-shift of 6 h counter-clockwise, and roughly 180° reversal of initial orientation with a clock-shift of 12 h, in agreement with the results obtained with homing pigeons (Schmidt-Koenig, 1958, 1960, 1961). Clock-shifts had no effect on nocturnal "nonsense" orientation. As later confirmed by Wallraff (1969) in laboratory conditioning experiments with mallards, the star compass (unlike the sun compass) is not time-compensated but operates on the basis of memorized star patterns.

Other experiments have dealt with the use of olfactory cues. A variety of evidence has been presented by Grubb (1971, 1972) that some petrels (*Oceanodroma* species) and some shearwaters (*Puffinus* species) may rely on olfactory cues to reach their burrows. In contrast to the controls, birds who had had their nostrils plugged, or who had sectioned olfactory nerves, were not home after one week. However, dealing with a different species, the wedge-tailed shearwater (*Puffinus pacificus*) in Hawaii, Shallenberger (1973) found little evidence for the use of olfactory cues.

Another set of experiments falling into the category of direct experimental interference is the work of Fiaschi *et al.* (1974) on swifts. Extending similar experiments of the Papi group on pigeons (Chapter V, Section E) to wild birds, Fiaschi *et al.* (1974) sectioned one olfactory nerve in 46 swifts and plugged one nostril with wax (ipsilaterally in control birds and contralaterally in experimental birds) to deprive them of olfactory information. When released 47–66 km away from their nests only three experimental birds returned, all of whom had lost their nose plugs, as against 15 control birds all of whom had their nose plugs in place. This result supports the view that olfactory cues may also play an important role in homing in other wild birds such as the Apodidae.

Summarizing the present state of experiments on the use of olfactory cues in wild birds, one finds evidence as inconsistent as that achieved with pigeons (Chapter V, Section E). Wild birds are difficult to handle in experimental field work, certainly much more difficult than the domestic homing pigeon. The section on experimental field work in wild birds is, therefore, comparatively short, and the following section on experimental field work with homing pigeons is consequently much longer.

V.   Homing experiments with pigeons

The homing pigeon is an excellent alternative to wild birds when studying bird orientation, as pigeons navigate in the true sense of the word. They
can be maintained in large numbers to produce large sample sizes, they
home (with some restrictions) throughout the year and they tolerate experimental interferences which wild birds would not tolerate. A lot of the
work on avian orientation and navigation, therefore, centres on elucidat-
ing the homing mechanisms of the domestic pigeon. So much data and
discussion has been accumulated during the last 10 years that the following account has to be selective.

   The homing pigeon is a domestic animal derived from the rock pigeon
(*Columba livia*) of the Mediterranean population and it was specifically
selected for fast and reliable homing. Comparing the homing abilities of
rock pigeons with that of homing pigeons, Alleva *et al.* (1975) found clear
superiority in the domestic birds over distances of more than 15 km.
This effect of selection is apparently not only due to superior physical
strength or aerodynamic ability, but also to enhanced central nervous
capacities reflecting higher sensory capabilities. Haase *et al.* (1977) found
a 5% higher allometric brain weight in homing pigeons than in non-homing fantails and strassers. The authors point out that the exact significance
of the larger brain of homing pigeons has yet to be clarified; it is not
necessarily an indication of greater learning ability. If some specific
structure of the central nervous system (CNS) should turn out to be particularly enlarged, it may offer a clue as to which sensory capabilities are
mostly involved in homing. In any event, this finding is by itself important because it is the first example in which domestication is paralleled
by an increase in brain weight. Feral pigeons ("commons"), descendants
of escaped domestic breeds (mostly homing pigeons), seem to hold an
intermediate position between homers and the rock pigeon, according to
Chelazzi and Pineschi (1974) and Edrich and Keeton (1977).

   Initial orientation of commons was indistinguishable from that of homers but homing speeds and homing success was considerably inferior to that of homers, possibly due to differences in social behaviour (e.g. a stronger drive to join other pigeons) rather than inferior orientational abilities. The sun compass was found to operate alike in both strains.

A. The sun compass

Even though the relevant capabilities have been demonstrated in laboratory experiments, the actual use of the sun compass or the magnetic compass cannot be taken for granted without specific confirmation in homing experiments. The mode of operation of the sun compass is well-established. Analogous to results obtained in conditioning experiments (Chapter III, Section B, 2), clock-shifts deflect the direction of initial orientation in homing experiments. Figure 45 A–C presents a summary of data involving 6 h and 12 h shifts (Schmidt-Koenig, 1958, 1961, 1969. 1972). Smaller degress of shift produce smaller deflections. In Fig. 46 sample sizes of around 100 experimental birds (shifted 2 h) and controls did not reach significance, with an actual mean difference of 15°. Homing performance was, however, significantly different. The effect of clock-shifts has been confirmed many times (e.g. Keeton, 1969; Walcott and Michener, 1971; Walcott, 1972) and is accepted as demonstrating the use of the sun compass, at least for initial orientation in pigeon homing. The equally drastic effect on homing speeds may indicate that the sun compass is also used *en route*, possibly for corrections that are, from other evidence, known to be performed.

For more than a decade the clock-shift experiment remained the only experimental interaction with pigeon homing to produce reliable and predictable results. Clock-shifts were widely used as experimental tools and they turned out to deflect birds at distances of several hundred kilometres, as well as within less than 2 km from the loft (Graue, 1963; Schmidt-Koenig, 1965; Alexander, 1975). Figure 47 adds some unpublished data of Schmidt-Koenig (cf 1971) to the published data, confirming that even with the loft building plainly visible to the human observer, at least some clock-shifted birds flew according to their shifted compass and some were even reported far away. If the loft was not visible (Graue, 1963; Alexander, 1975) more birds were led astray. These findings are important inasmuch as they demonstrate that pigeons try to navigate even when close to the loft and that they pay little if any attention to landmarks.

Additional experiments involving clock-shifts provided further insight into this. After a great deal of inappropriate experimentation and discus-

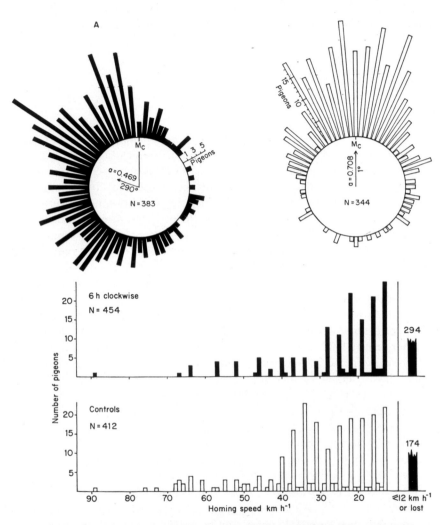

Fig. 45. Summary of initial orientation (circular diagrams) and homing performance (rectangular diagrams) in the clock-shifting experiments of Schmidt-Koenig (1961, 1969, 1972). Experimental birds are given in solid bars, controls in open bars. The time shift involved in series A, B, C is indicated in each series. The length of bars are proportional to the number of birds according to the scale given at each diagram. Initial orientation of experimental birds is plotted with reference to the mean of controls (M_C)

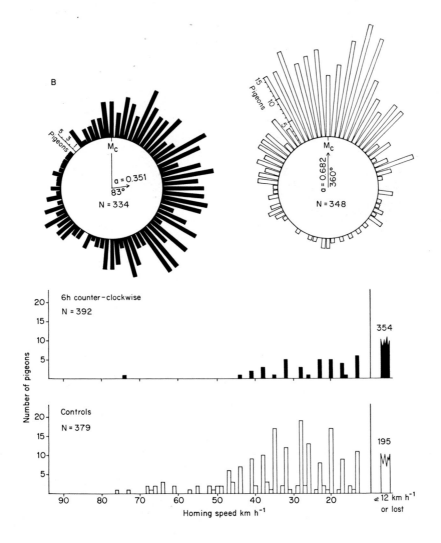

B

5, 3
Pigeons

M_C

$a = 0.351$
83°

N = 334

15
10
Pigeons
5

M_C

$a = 0.682$
360°

N = 348

6h counter-clockwise
N = 392

354

Number of pigeons

20
15
10
5

Controls
N = 379

20
15
10
5

195

90   80   70   60   50   40   30   20   ≤ 12 km h⁻¹
or lost

Homing speed km h⁻¹

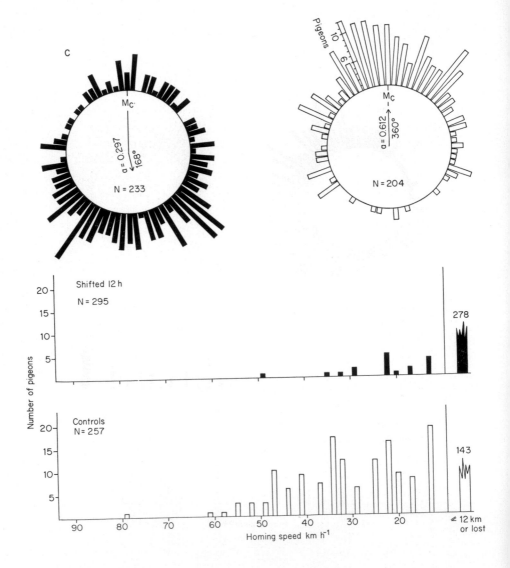

C

$M_{C'}$

$a = 0.297$
$168°$

N = 233

pigeons
10
6

M_{C}

$a = 0.612$
$360°$

N = 204

Shifted 12 h
N = 295

20
15
10
5

278

Number of pigeons

Controls
N = 257

20
15
10
5

143

90   80   70   60   50   40   30   20   ≰ 12 km
                                                    or lost

Homing speed km h⁻¹

sion on the effect of overcast (Matthews, 1953b; Wallraff, 1960a, 1966b; for details see Schmidt-Koenig, 1965), Keeton (1969, 1974a), who was fortunate in having prolonged periods of total overcast at Ithaca, New York, found that not only normal birds were homeward-orientated, but that even clock-shifted experimental birds performed like controls under overcast (see Fig. 48). Two conclusions are possible from these results:

(1) the sun is indeed used as a compass when available and;

(2) if the sun is unavailable, alternative mechanisms which are not subject to clock-shifts, replace the sun compass.

The question of how the sun compass is established ontogenetically is discussed in Chapter V, Section B, 1. Related to this question, Edrich and Keeton (1978), extended experiments by Alexander and Keeton (1974) and Keeton and Alexander (1978) which had indicated that the degree of deflection accomplished by clock-shifts is a function of how much and

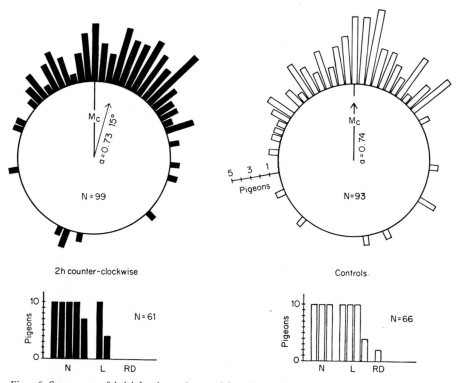

Fig. 46. Summary of initial orientation and homing performance of controls and experimentals shifted 2 h counter-clockwise in releases from 460, 320 and 280 km. In the rectangular diagrams RD symbolizes birds home during the release day, L birds that homed later and N birds that never homed. All other symbols as in Fig. 37. From Schmidt-Koenig (1972).

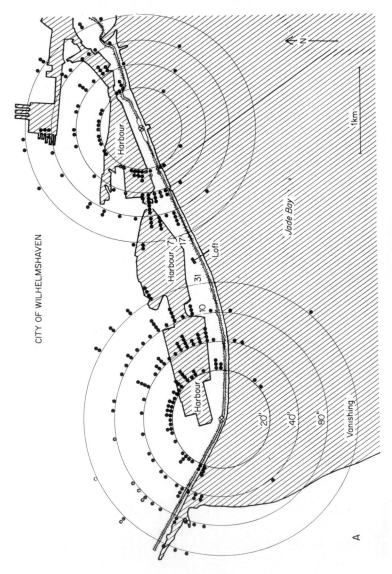

CITY OF WILHELMSHAVEN

Fig. 47. Initial orientation (A) of pigeons shifted 6 h clockwise released 1800 m west of the loft, and of pigeons shifted 6 h counter-clockwise released 1650 m east of the loft; and (B) of corresponding control birds superimposed on a map showing obvious topographical features. Bearings 20 s, 40 s, 80 s after release and at vanishing, are summarized from a total of five releases, three from the west and two from the east. Figures are given instead of solid (experimental) or open (control) symbols at those bearings at which more than five birds were scored. The loft was at the limit of visibility of a vanishing pigeon. Unpublished data of Schmidt-Koenig.

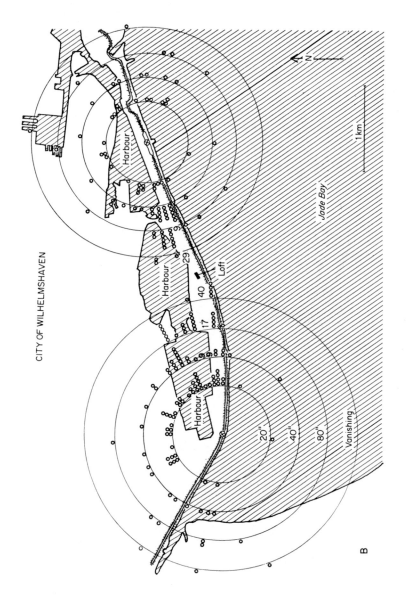

CITY OF WILHELMSHAVEN

Harbour

Harbour

Harbour

Loft

Jade Bay

N

1 km

20"
40"
80"
Vanishing

B

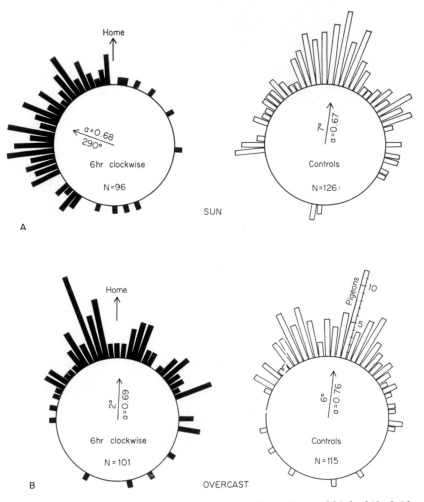

Fig. 48. Summary of initial orientation of control and experimental birds shifted 6 h clockwise, A, under sun and B, under overcast. Symbols as in Fig. 37. Modified from Keeton (1974a).

what kind of free flight is available to the birds during the shifting process. If, during the common light-period of the shifted and the natural day, the birds were allowed to see the sun from an aviary without free flight, full deflection, i.e. roughly 90° for 6 h of shift was accomplished. Watching the sun from an aviary does not suffice to enable the bird to recognize that the clock and sun are not in agreement, although differences may be drastic, not only if azimuth positions were compared with landmarks, but also with sun altitudes and falling rather than rising movements.

Free flight around the loft or exercise flights in large aviaries reduced the degree of deflection, indicating that some other reference system available only in flight is used by the birds. The reduction was greater when the birds were made to fly in aviaries aligned east–west, and smalled in aviaries aligned north–south. This may indicate that the earth's magnetic field has its share in aligning the sun compass. The results support findings (Chapter V, Section B, 1) that the sun compass may be recalibrated and is used, once it is established, when the sun is available.

Homing without the sun has been demonstrated in nocturnal pigeon races (cf Levi, 1963) and in experiments (St. Paul, 1962; Keeton, 1970b; Goodloe, 1974), and there is increasing radar evidence that migrating birds continued to fly even within clouds where neither celestial nor terrestial cues were visible (e.g. Bellrose and Graber, 1963: Williams *et al.*, 1972; Griffin, 1972, 1973).

B. Magnetic cues

After Wiltschko's laboratory findings of magnetic orientation in European robins and warblers (Chapter III, Section B, 4) and magnetic effects on other animals, e.g. insects (Becker, 1963; Picton, 1966; Lindauer and Martin, 1968, 1972, 1977), the situation was sufficiently encouraging to resume previously unsuccessful attempts to look for a magnetic alternative to the sun compass in pigeons. Homing experiments in which bar magnets were unsuccessfully attached to pigeons failed to produce significant disorientation, e.g. when carried out by Gordon (1948), Matthews (1951), Riper and Kalmbach (1952), Yeagley (1951) and Schmidt-Koenig (unpublished data), with only one exception (Yeagley, 1947) who was also the first to correlate sun spot activity with the homing speed of pigeons. All experiments had been carried out in sunshine. After finding no difference between experimental birds and controls in the sun, Keeton (1971a, 1972) finally achieved disorientation in experienced birds carrying bar magnets under overcast, and in inexperienced birds even in the sun, as shown in Fig. 49.

Walcott (1972) designed Helmholtz coils, one each glued on the head and around the neck of a pigeon, and a battery on its back. Current flowing through the coils produced a magnetic field of around 1 Oe between the coils at the pigeon's head. In an initial series of experiments (Walcott, 1972) found that experimental birds with current

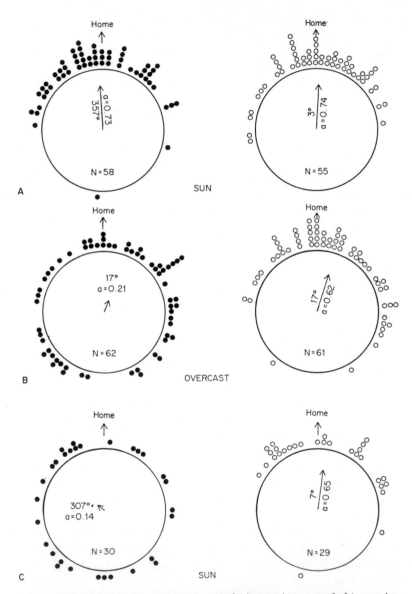

Fig. 49. Summarized initial orientation of control pigeons (open symbols) carrying brass bars and experimental pigeons carrying magnet bars; A, experienced birds in the sun, B, experienced birds under overcast, C, inexperienced first-flight birds in the sun. From Keeton (1972, 1974a).

switched on were more scattered than controls with disconnected batteries. In a subsequent series of experiments carried out by Walcott and Green (1974), experimental and control birds had their batteries connected, but in one group the current flow was clockwise and in the other it was counter-clockwise (both at o·6 Oe). As shown in Fig. 50, birds with the "north-up" arrangement were not oriented towards home under overcast. These findings probably link up with Wiltschko's concept of the inclination compass.

In a subsequent series of homing experiments Walcott (1977) tested the effect of low intensity artificial fields, again applied by Helmholtz coils. The 16 km radio bearings (not directly comparable to the field-glass vanishing bearings usually recorded by others) produced by the birds were more scattered in the case of the experimental birds when o·1 Oe fields were applied, the same as the controls when fields of o·3 Oe were applied, and again somewhat different when a o·6 Oe field (about equal to the geomagnetic field) was applied. Since all releases were made in the sun, Walcott interprets his results to indicate that the pigeons did not simply switch between their magnetic and their sun compass, but that the systems interact.

If the geomagnetic field is used for orientation, the question arises how birds deal with natural magnetic anomalies, and this question has been considered by Walcott (1978). All researchers working with homing pigeons have described release sites or extended areas where birds, when released, have obvious difficulties in selecting the home direction or any direction at all, and the homing performance is usually poor. A magnetic anomaly south-south-west of Walcott's loft near Boston, Massachusetts, is shown in Fig. 51. Walcott released a number of pigeons in the sunlight, within this area, and radio-tracked them within a range of roughly 25km. The resulting jumble of tracks (Fig. 52) explains itself. Tracks straightened only outside the anomalous area. This is a particularly clear example of the interaction of the geomagnetic field and initial orientation. However, not all locations persistently plagued with similar patterns of disorientation can be explained by anomalies of the geomagnetic field. The Jersey Hill Firetower release site 120 km west of Ithaca, New York, and the adjacent area as described by Keeton (1970a, 1971b), is such an example. Initial orientation and homing is consistently very bad, but geomagnetic maps do not reveal any anomalies.

Apparently, magnetic sensitivity in pigeons is so high that even normal fluctuations of the magnetic field of the earth could be shown to influence pigeons in otherwise unmanipulated homing flights. Keeton *et al.* (1974a,b) released the same group of similarly aged, experienced homing pigeons from one particular release site near Weedsport, New York, 73 km north of the loft, repeatedly throughout the season, in the sunlight and without

D

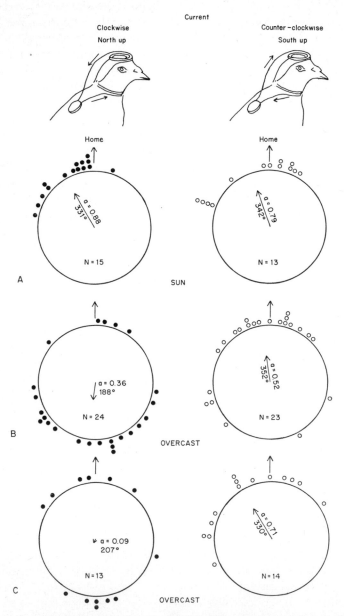

Fig. 50. Summarized distribution at 15 km distance from the release site of radio-tracked pigeons wearing Helmholtz coils with current flow clockwise, i.e. north up (black symbols) and with current flow counter-clockwise, i.e. south up (open symbols). A, in the sun; B, under overcast, birds released for the first time at that site; C, as in B, but treatment reversed among groups and birds released for the second time from that site. Fom Wallcott and Green (1974).

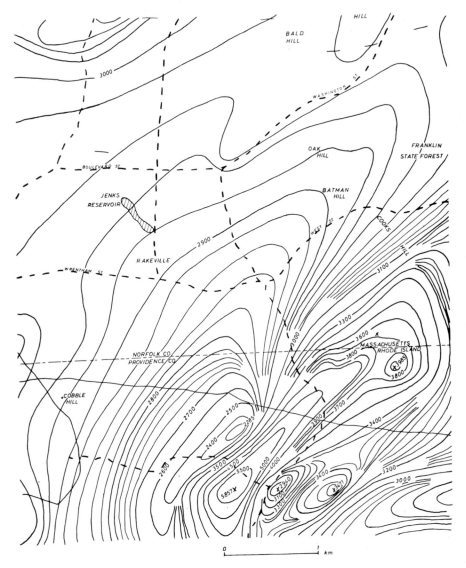

Fig. 51. Magnetic anomaly near Providence, Rhode Island, where the tracks of Fig. 52 were recorded. From Wallcott (1978).

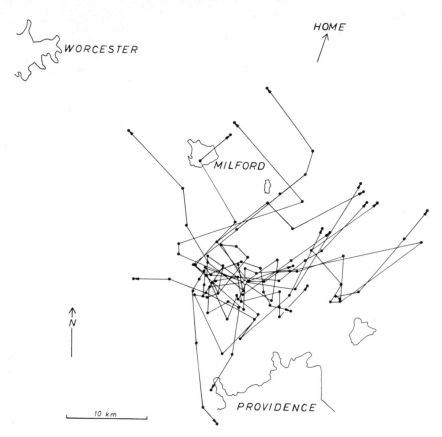

Fig. 52. Radio tracks of pigeons released by Wallcott (1978) in the magnetic anomaly shown in Fig. 51.

any intermediate releases from elsewhere. This preparation eliminated the variability of performance that is usually encountered and is attributed to some special properties of the release sites, the direction of release the individual differences of the birds etc. (cf Schmidt-Koenig, 1965; Keeton, 1974a), to such a degree that small but significant changes in initial orientation emerged that can be correlated with fluctuations of the earth's magnetic field.

As shown in Fig. 53, initial orientation is normally characterized by a deviation of about 20° to the right of home, shifted more to the home direction with increasing magnetic activity. But why should orientation towards home improve with increasing disturbance? The answer to this question emerged when a similar series of releases was performed from Campbell, New York, a site 70 km west-south-west of home. This site is

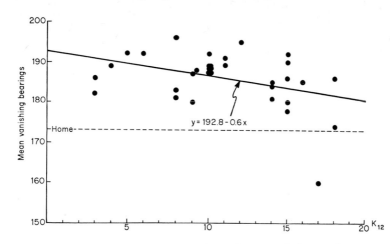

Fig. 53. Mean vanishing bearings (ordinate) of 33 releases made in 1973 from Weedsport, New York, plotted as a function of K_{12} (abscissa) and the resulting regression line. K_{12} is the sum of the four K values (for magnetic disturbances obtained from the magnetic observatory at Fredericsburg, Virginia) for the 12-hour period ending at the time the release was completed. From Keeton *et al.* (1974).

normally characterized by a bias of about 15° to the left of the home direction. Here increasing magnetic activity shifted the initial orientation of the birds away from home. Comparing the results from Weedsport and from Campbell in Fig. 54, increasing magnetic activity can be concluded to result in a counter-clockwise shift of initial orientation, without any reference to the home direction.

In additional experiments Larkin and Keeton (1976) checked the initial orientation of experimental birds at Weedsport, who had bar magnets glued to their backs, against that of control birds carrying brass bars. Figure 55 shows the result: the mean deviation from the home direction of control birds continued to be inversely correlated with the K_{12} values. The coefficient of correlation was calculated to be -0.403 and to be significant. The mean deviation from the home direction of experimental birds turned out not to be correlated with the variations in K_{12} values, the coefficient of correlation attaining a value of $+0.38$ and not to be significant. These results strongly support the interpretation that there is a direct cause-and-effect relationship between the fluctuations in the earth's magnetic field and the variations in the initial orientation of pigeons.

These results and those reported earlier confirm that pigeons are more sensitive than expected to small changes of magnetic intensity. Despite this valuable information, the experimental outcome does as yet neither explain why increased magnetic disturbance gives rise to a counter-clockwise shift of initial orientation, nor does it seem to provide any further

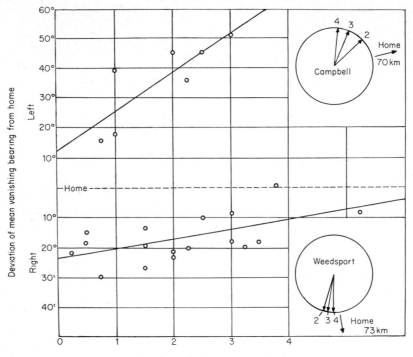

Fig. 54. Effect of normal fluctuations of the earth's magnetic field on initial orientation of homing pigeons in relation to the home direction at Weedsport, New York (lower half) and Campbell New York (upper half). In the rectangular graph, mean vanishing bearings are plotted as deviations from the home direction (ordinate) with reference to K values (cf. legend of Fig. 53). After Keeton (1974b).

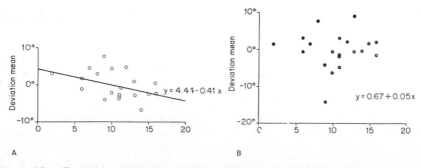

A   B

Fig. 55. The effect of bar magnets on initial orientation at Weedsport. Deviation means (ordinate) plotted as a function of K_{12} values (abscissa). A, control birds (open symbols) carrying brass bars continued to show the effect as given in Fig. 53 (different scale). B, experimental birds (black symbols) carrying bar magnets. The least square regression line is not shown; it is nearly horizontal and virtually meaningless. From Larkin and Keeton (1976).

clue as to how magnetic fields are sensed by birds. It remains an open question as to whether the observed effect is one involving the magnetic compass, or whether the earth's magnetic field also provides navigational information, as suggested by the outcome of experiments by Kiepenheuer (1978a) and by Wiltschko and Wiltschko (1978) discussed below.

1. Hierarchy of compasses

Before turning to evidence indicating that the earth's magnetic field may not only be used for compass orientation, but may also provide navigational information, I should like to discuss briefly the question of the hierarchy of compasses in pigeon homing, as discussed for migratory birds in Chapter III, Section B, 5.

Wiltschko *et al.* (1976) reared pigeons in an artificial photoperiod which was constantly shifted 6 h counter-clockwise. The birds were allowed to see the sun during their physiological morning that was really the afternoon. On release no deflection of initial orientation 90° clockwise was observed, as has been the case in so many releases before (Chapter V, Section A). There was no difference to control birds. Later exposure to a normal day initially produced a counter-clockwise deflection of initial orientation, which was normally seen after a clockwise shift. After some time in natural conditions the deflection of initial orientation disappeared again. Thus, young pigeons seemed to calibrate their sun compass by a learning process and the sun compass could be recalibrated more than once, as shown by further experiments. By analogy with the results regarding European robins (Chapter III, Section B, 5), the sun compass may well be calibrated relative to the magnetic compass.

2. Navigational information from the geomagnetic field

Wiltschko and Wiltschko (1978) used a Volkswagen squareback to transport pigeons to the release site, and disturbed orientation was sometimes observed after the pigeons' crate had travelled on top of the engine, with magnetic activity produced by the generator. Walcott (pers. comm.) later confirmed this "VW-effect".

Kiepenheuer (1978b) and Wiltschko and Wiltschko (1978) approached the question of magnetic information obtained *en route* to the release site systematically and methodologically in several different ways. Kiepenheuer (1978b) transported his experimental birds in a crate which was inside

Helmholtz coils, on top of a van. The coils reversed the vertical component of the earth's field (condition C in Fig. 33). According to the model of an inclination compass this should reverse the pigeons' compass. Changes in the car's direction during the outward journey to the release site do not alter these conditions. As may be seen from Fig. 56 the initial orientation of experimental birds was deflected to the right.

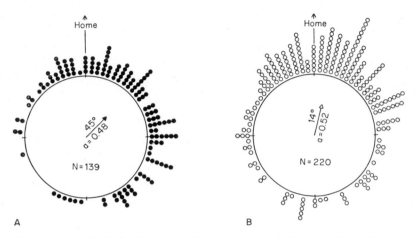

Fig. 56. Summary of initial orientation of, A, experimental pigeons subjected to a reversed vertical magnetic field during displacement, B, control pigeons exposed to the normal geomagnetic field during displacement. From Kiepenheuer (1978b).

In a second series of experiments Kiepenheuer (1978d) reversed the horizontal component (condition illustrated in Fig. 33D) during transport, also reversing the proposed inclination compass plus reversing polarity. During transport to the release site Kiepenheuer compensated for all the turns by keeping the coils and the crate always oriented in their initial orientation, with respect to the geomagnetic field, by a wheel and belt arrangement. Figure 57 shows the result. The summarized initial orientation of experimental birds (Fig. 57E) shows a tendency towards bimodality and this may be explained as a partial reversal of orientation. Surprisingly, however, controls without coils but whose crate was also kept constant by compensatory rotation, showed the same effect (Fig. 57C). Wiltschko and Wiltschko (1978) also reversed the horizontal component during transport inside Helmholtz coils (condition D in Fig. 33). The turns of the car during the outward journey do alter these artificial conditions, and Wiltschko therefore used a route for displacement that deviated no more than 25° from a straight line to the south-south-west. Initial orientation of experimental birds was not different from random. Papi *et al.*

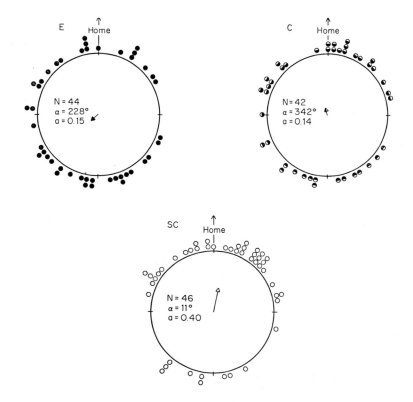

Fig. 57. Preliminary account of the initial orientation of pigeons deprived of information concerning turns during displacement. Six releases have so far been carried out.

E Same as C but with the horizontal component of the magnetic field artificially reversed.

C Pigeon crate kept fixed with respect to magnetic north during transport to the release site on top of the car.

SC Supercontrols, carried in crates fixed to the top of the car as usual.

(1978) displaced experimental pigeons in iron containers and control birds in aluminium containers. Though additional manipulation of air supply—bottled air or natural air—complicated the interpretation, some effect could be attributed to the magnetic field being altered in the iron container.

Thus pigeons do seem to pay attention to the geomagnetic field during displacement: they sample information *en route*, and the geomagnetic field is not used only for compass orientation. In other words, we are dealing with a "compass" that is not precisely what one ordinarily calls a compass, but a system with navigational capacities. The simplest model would be a system of route reversal: directions recorded by a magnetic compass during the outward journey could be integrated over time to provide the home direction at the release site. A very similar mechanism of route reversal is known to operate in bees (von Frisch, 1968) involving the sun compass.

C. Gravitation

The magnetic inclination compass as proposed by Wiltschko (Chapter III, Section B, 4) would require reference to the vertical, probably measured in terms of gravity. In some instances direct interaction of magnetic fields and the influence of gravity has been clearly demonstrated. The error committed by bees dancing on the vertical comb to communicate the direction of food to hive mates, is a case in point (Lindauer and Martin, 1968, 1972; Martin and Lindauer, 1977). There is no error ("Missweisung") if the direction of the dance (orientated to gravity) and the direction of magnetic field lines coincide. Errors are largest if the direction of the magnetic field lines and the direction of the dance diverges most.

Keeton and Larkin (1978) speculated that the correlation between the fluctuations of the geomagnetic field and initial orientation, as discussed in Chapter V, Section B, involves gravity. The changing spatial relationship between the earth, the sun and the moon causes traceable variations in gravity on earth and might measurably influence initial orientation. In similar experiments on magnetic effects, one group of pigeons was released repeatedly from one of a total of three release sites. The variation of the mean vanishing bearings was then related to the day of the lunar month. A total of six series of experiments was performed in the course of four years. A significant relationship between mean vanishing bearings and the day of the lunar month emerged. In some years mean vanishing bearings had a cyclical relationship in phase with the day of the lunar month measured from new moon to new moon, and in other years the cycle was from full moon to full moon.

Similar relationships (seemingly stable in one of two modes 180° out of phase) have previously been found in monthly tidal rhythms in invertebrates (e.g. Enright, 1972), but a direct relationship between gravity and orientation has not emerged. Mean vanishing bearing revolves once per lunar month and gravity revolves twice per lunar month. There might be an indirect relationship between these two which has yet to be demonstated.

D. Visual cues

If an animal shows some outstanding performance and one sets out to find the sensory basis of this performance, the researcher is usually well-advised to look for a correspondingly well-developed sense organ. Homing is such an outstanding performance and the eye is the predominant sense organ in the head of pigeons and many other birds so it is only natural that the eye and vision have been held responsible for homing ability. High visual ability has been assumed and extrapolated from anatomical data of the retina and the dioptric apparatus, but corresponding experimental evidence (especially of birds in flight) is not yet available.

The hypothesis of the central role of vision in pigeon homing collapsed

Fig. 58. Pigeon equipped with frosted lenses on a ring of velcro. From Schmidt-Koenig.

when Schlichte and Schmidt-Koenig (1971), Schmidt-Koenig and Schlichte (1972) and Schlichte (1971, 1973) reduced vision to near blindness by inserting frosted contact lenses into pigeons' eyes. This simple device permitted experimenting in flight and during actual homing processes. The original contact lenses were later replaced by cups attached to the area around the eye on a ring of velcro (Fig. 58) rather than fitted directly onto the cornea.

Cardiac conditioning experiments (Chapter III, Section B, 1) had shown that pigeons fitted with lenses were able to recognize a red pole 5 cm in diameter when 2 m away, but were unable to recognize either the pole or more complex structures, such as the entrance of the loft, when 6 m or more away. On release some birds demonstrated their visual impairment by hitting wires, trees and buildings. Nevertheless, initial orientation was no different from that of controls, as may be seen in Fig. 59 (circular diagrams).

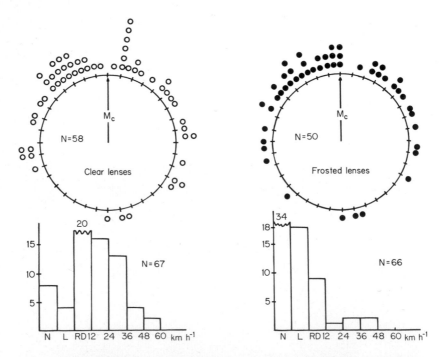

Fig. 59. Summarized initial orientation (circular diagrams) and homing performance (histograms) of control birds with clear lenses (left) and of experimental birds with frosted lenses (right). In the circular diagrams each symbol represents the vanishing bearing of one bird plotted with reference to the mean of controls (M_c). Homing performance is summarized (from right to left) in classes of 12 km h^{-1} down to 12 km h^{-1}; RD summarizes all birds home on the day of release but with slower speed than 12 km h^{-1}; L summarizes all birds that homed later and N summarizes birds that did not home at all. From Schmidt-Koenig and Schlichte (1972).

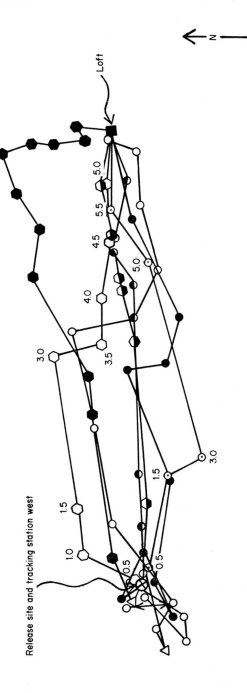

Tracking station north

Release site and tracking station west

Loft

N

1 km

1.0 1.5 3.0 3.5 4.0 4.5 5.5 5.0

0.5 0.5 1.5 5.0 5.5

3.0

Fig. 60. Approximate tracks of seven control birds (different symbols) wearing clear lenses. The figures indicate time in minutes elapsed since take-off. From Schmidt-Koenig, Hartwick and Kiepenheuer, unpublished data.

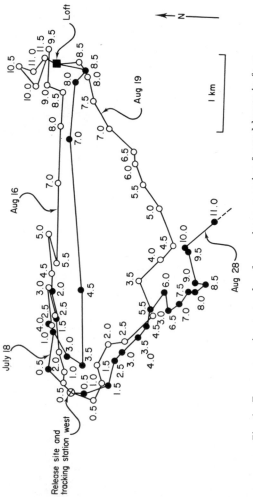

Tracking station north

Fig. 61. Four approximate tracks of one pigeon wearing frosted lenses in four consecutive releases, from the same site as in Fig. 60. Some tracks are shown in black symbols for easier discrimination. Figures indicate minutes elapsed since take-off. Unpublished data of Schmidt-Koenig, Hartwick and Kiepenheuer.

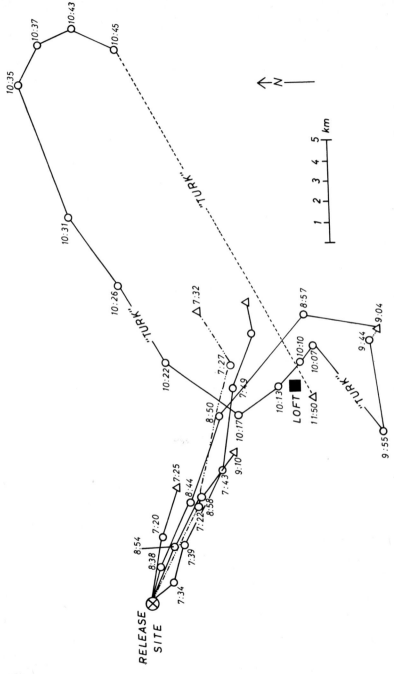

Fig. 62. Approximate tracks of five experimental birds wearing frosted lenses released 15 km NW of the loft. Figures indicate time of day. Triangles symbolize location where a bird perched. The dashed line of the pigeon "Turk" connects locations and times between which radio contact was temporarily lost. From Schmidt-Koenig and Walcott (1973).

The homing performance of these birds, was, however, different in that (see Fig. 58) few experimental birds homed as fast as average controls, more experimental birds than controls homed during the days following the release and more experimental birds were lost. Some of those experimental birds that did home hovered down directly into the loft, while others landed nearby and tried another approach or were retrieved. Thus vision, certainly image vision, did not seem to be necessary to head for home from a distance. Why did experimental birds not arrive home in the same numbers as the controls? Radio tracking from an aeroplane or from the ground (Schmidt-Koenig and Walcott, 1973, 1978) provided some answers. Figure 60 shows the approximate tracks of seven control birds wearing clear lenses: after some initial circling they flew home without much of a detour. Figure 61 presents four approximate tracks of one outstanding pigeon that homed directly three out of four times when fitted with frosted lenses. Figure 62 presents less exceptional results: five birds were released 15 km north-west of the loft, and initial orientation was, as usual, towards home. Two birds perched half-way between the release site and home and two birds gave up just after they had passed the loft. "Turk" tried hard but missed the loft several times and was finally tracked down and picked up less than 1 km from the loft.

Extrapolating from these and numerous other similar results of tracking experiments, pigeons with drastically reduced vision begin to have difficulties when trying to pin-point the loft. During the final approach, visual cues seem to be important although many birds finally did find the loft after repeated attempts during the days following the release day. Thus, for the navigational part of the homing flight, i.e. determining which direction is the home direction, visual cues turned out not to be essential. This navigation system is largely non-visual and guides the pigeon with amazing accuracy to the vicinity of the loft. The birds also seem to know when they are home and even when they have missed the loft and distance is again increasing. The sun compass may still operate, as reported in conditioning experiments by Schlichte (1973) and in homing experiments by Schmidt-Koenig and Keeton (1977); its operation was demonstrated in the classical manner: with frosted lenses in place, clock-shifted birds were deflected as expected and as shown previously in birds without lenses. The evidence that the sun compass is still used is no argument against non-visual navigation since the sun is not used for navigation.

Haase *et al.* (1977) comparing eye (and brain) weights in homing and non-homing breeds of pigeons found no allometric differences. This suggests that homing ability is not associated with good vision and supports the finding that accurate vision is not essential for homing.

E. Olfactory cues

With the importance of visual cues downgraded and magnetism by itself insufficient to explain homing, the search for other non-visual cues involved in pigeon homing was revived. Despite the low classification of the sense of smell in most birds including pigeons, Papi and his co-workers gave it a try and performed a number of well-designed homing experiments. Papi *et al.* (1971, 1972) and Benvenuti *et al.* (1973a) released birds whose olfactory nerves had been sectioned or whose nostrils had been plugged with cotton or modelling clay (see also Snyder and Cheney, 1975). Experimental birds performed more poorly in initial orientation and in homing performance than controls. From these results Papi *et al.* (1972) proposed that pigeons at their home loft associate odorous substances carried by winds with the direction of those winds. If displaced and released the bird would recognize the local odour, reverse the direction of the wind carrying it to the loft and fly home. Subsequent experiments seemed to confirm this hypothesis. Papi *et al.* (1974) raised two groups of pigeons in aviaries (Fig. 63) walled with plastic and bamboo, which permitted air to enter diffusely. Both groups of birds had access to one of two glass-walled corridors on top of the aviary. Repeatedly, air was blown through the corridors, alternating

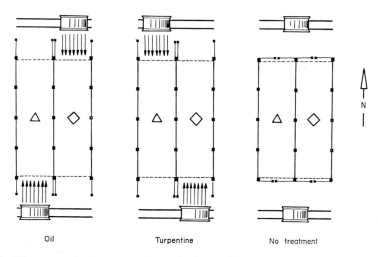

Oil   Turpentine   No treatment

Fig. 63. Diagrammatic view from above of the corridor-aviaries with fans blowing air through the corridors. The triangles and diamonds are referred to in the text. From Papi *et al.* (1974).

for one group (triangles in Fig. 63) with the volatile components of olive oil added from the south and with synthetic turpentine added from north. The other group received the same treatment reversed. At the release site a drop of olive oil was applied onto the beak of half the experimental birds, and some turpentine was placed on the beaks of the other half. In releases from easterly directions the oil-triangle and the turpentine-diamond birds headed by and large in southerly directions, while the oil-diamond and the turpentine-triangle birds headed mostly in northerly directions, which was as expected. Individual homing speeds were not different among the groups. The fact that many of the birds did home does not clearly support the authors' hypothesis, as they admit.

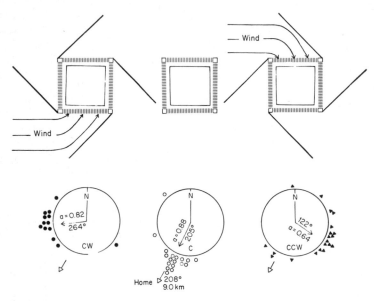

Fig. 64. The deflector aviary experiment, upper row: aviaries with deflectors and control aviary without deflectors. Initial orientation as obtained in the first release is given in the circular diagram below each aviary. CW, birds expected to deviate clockwise; C, control birds; CCW, birds expected to deviate counter-clockwise. From Baldaccini *et al.* (1975).

Baldaccini *et al.* (1975) raised pigeons in aviaries equipped with deflectors to deflect the winds (Fig. 64), thereby producing an association of odorous substances with false directions. Predictably, deflected initial orientation was recorded as shown in Fig. 64. Homing performance was not convincingly different.

Papi soon modified his initial hypothesis (Papi *et al.*, 1972) that pigeons recognize, at the release site, odorous substances which are familiar to

them from the winds at the loft. Detour experiments provided evidence
that olfactory cues are also sampled during displacement (Papi *et al.*, 1973;
Papi, 1976), informing the bird of the direction of displacement. In these
detour experiments two groups of birds were transported along routes
that left home in opposite directions, for example to the north-west and
to the south-east, to finally converge at a release site south-west of the loft
(Fig. 65A). The group initially transported to the north-west would sample

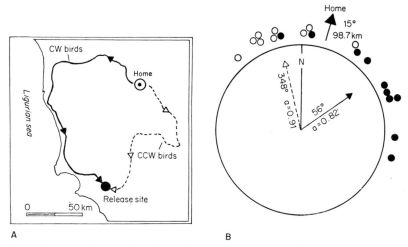

Fig. 65. A, design and, B, result of one detour experiment. Broken detour line, broken
mean vector and open symbols designate CCW birds; solid line, solid mean vector
and solid symbols designate CW birds. For more methodological details see text.
From Papi (1976).

odours *en route* (as long as they were familiar with them through the
winds at the loft) and realize a north-west displacement had taken place
and head into easterly directions (deflected clock-wise), despite actually
being transported to the south-west finally through olfactorily unfamiliar
territory. The other group would interpret its initial route correspondingly
to be a south-east displacement and head into north-westerly directions, or
be deflected counter-clockwise from the home direction. As may be seen
from Fig. 65B birds performed according to expectation. In most other
detour experiments and in a repetition by Fiaschi and Wagner (1976)
similar results were obtained.

In additional experiments Papi (1976) modified the detour experiment. In
order to prevent sampling of olfactory stimuli *en route*, birds were displaced with their nostrils plugged or in air-tight containers. As judged from
the outcome of some preliminary results, initial orientation was affected.

Hermayer and Keeton (1979, in prep.) repeated the olfactory nerve sectioning experiment. The results were quite similar to those reported by Papi *et al.* (1971) and Benvenuti *et al.* (1973), but Hermayer and Keeton were not convinced that these experiments had dealt with a primary factor in homing. There was less difference between initial orientation of experimental birds and controls than in homing speeds. Hermayer and Keeton suggested and Papi *et al.* (1979) in a later repetition of this experiment confirmed by additional radio tracking, that experimental birds remained stationary for extended periods on the way home. In order to avoid trauma and other non-olfactory effects of surgical interaction. Keeton *et al.* (1977) inserted small plastic tubes through the entire length of the nasal passage forming an air passage between the nostrils and the epiglottal region, so that air completely bypassed the olfactory epithelium. In five releases from unfamiliar sites less than 25 km away, experimental and control birds (i.e. wearing only one tube) performed no differently. In 11 releases from unfamiliar sites more than 76 km away, the initial orientation of experimental and control birds was not significantly different. As in the surgical experiments, however, the homing of experimental birds was much poorer than that of controls. In view of the undisturbed initial orientation, the authors again concluded that olfaction is not essential for homeward orientation.

Papi and co-workers quickly repeated this experiment in Italy (Hartwick *et al.*, 1977). From an unfamiliar site homing and initial orientation of experimental birds fitted with tubes was clearly inferior to that of controls. From a familiar site initial orientation and homing of experimental birds was significantly poorer than that of controls but the difference was not as drastic as in the former experiment.

Repeating Papi's detour experiment near Ithaca, New York, Keeton (1974c) found no evidence for the effect of detours on any of the customary criteria. He carried out 15 experiments using five different groups of birds, with differing homing experience, from three release sites. Hartwick *et al.* (1978) repeated the detour experiment in Germany. Only at one out of a dozen release sites (Marktheidenfeld-Glasofen, 70 km east of Frankfurt), was there a clear-cut and consistent deflection as in most of Papi's experiments in Italy. Only 2 km from that site no more effect could be recorded, and at the other sites results were inconclusive. Only an overall summary of initial orientation reaches significance, because of the results at Glasofen. Papi (1976) reinterpreted a preliminary account of the data of Hartwick *et al.* (1978) as supporting his results, but he ignored that the seemingly positive summary was based on data from the Glasofen site.

Other striking examples of diverging results are experiments in which strongly odorous substances such as α-pinene, olive oil and home-made plant extracts were applied on or near the nostrils of the birds. The idea was that these strong odours might work as masking agents preventing the birds from recognizing the subtle odours supposedly used for homing. Benvenuti *et al.* (1973b) and Fiaschi and Wagner (1976) reported positive results: initial orientation and homing of experimental birds was somewhat inferior to that of controls. Keeton and Brown (1976) were unable to find consistent differences between experimental birds and control birds in follow-up experiments. Papi (1976) reinterpreted Keeton and Brown's (1976) data claiming that the homing behaviour of experimental birds was disturbed, while Benvenuti *et al.* (1977) agreed with Keeton that previous data, including those from Italy, were not convincing. In a new series of modified experiments Benvenuti *et al.* (1977) again reported that homing behaviour was disturbed, but repetitions by Hartwick et al. (1978) in south-western Germany yielded inconsistent results. Together these results confirm a stronger emphasis on olfactory cues in Italy and their rather low order of magnitude in pigeon homing elsewhere.

The deflector loft experiment is, to date, the only one which has been repeated in which the results agree with the original (Kiepenheuer, 1978a). However, when Kiepenheuer (1979) eliminated olfaction by local anaesthesia of the birds' olfactory epithelium with xylocain, the deflection attributed to directionally altered olfactory information continued (Fig. 66). Thus, the deflector aviary result seems to require interpretations other than olfactory orientation. Prior to Kiepenheuer's findings local anaesthesia of the olfactory membrane (Schmidt-Koenig and Phillips, 1978) was used as a potent method in experiments on the role of olfaction and homing in the same laboratory. As discussed in Chapter III, Section A, 3 in cardiac conditioning experiments xylocain spray eliminated the response to low concentrations of amylacetate for about 90 min. This pharmacological approach has the advantage of being very effective in excluding olfactory information yet being at the same time completely reversible. In contrast to surgically severing the olfactory nerves and to inserting tubes into the nasal passage, local anaesthesia is only mildly traumatic to the bird and, if done properly, does not have obvious systemic effects.

In homing experiments Schmidt-Koenig and Phillips (1978) performed eight releases from release sites 40 km away. The experimental group was sprayed with xylocain immediately before displacement (XBD), one control group was sprayed before displacement with a preparation identical to the experimental spray but lacking xylocain (NXBD), and another control group was sprayed with xylocain upon arrival at the release site (XAR). Figure 67 presents the summarized initial orientation and homing performance. Initial orientation of all three groups was non-random ($P < 0.01$) and directed towards home ($P < 0.01$). There was no difference between any of the groups ($P > 0.05$). Likewise, homing performance was at perfectly normal levels without any significant differences ($P > 0.05$)

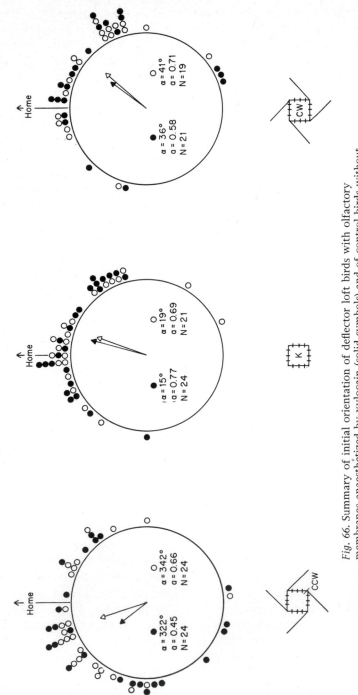

Fig. 66. Summary of initial orientation of deflector loft birds with olfactory membranes anaesthetized by xylocain (solid symbols) and of control birds without treatment. From Kiepenheuer (1978b).

among the groups. Treatment was rotated among the groups. It would, therefore, be useful to consider the first three releases in which birds of each group were receiving their first exposure to each treatment. Only then were there some marginally (P >0·01) significant differences between the experimental and the two control groups. Releases from 25 km and from 70 km involving repeated application of xylocain to eliminate olfaction during transport to the release site, the waiting period and beyond the time of release, likewise produced no difference in the initial orientation of experimental birds, non-xylocain controls and untreated controls (Schmidt-Koenig and Phillips, 1978; Schmidt-Koenig, 1978). There were only marginal and inconsistent differences in homing performance of experimental birds and untreated controls.

The effect of eliminating olfaction by local anaesthesia was at best slight and inconsistent and not as clear-cut as in similar experiments by Papi and co-workers, or as strong and consistent as, e.g. that obtained by clock-shifts. The birds did home without olfactory information during the outward journey at the release site and beyond the time of release. These results support the view shared by Keeton that olfactory information during displacement is not essential for successful initial orientation or homing. It is possible that we may be dealing not with olfactory cues but with something very similar.

The differing results obtained in Italy, the USA and Germany may well be caused by differential selection for different homing systems or by different emphasis on a number of different systems. Juvenile birds from Pisa (Italy) and from Tübingen (West Germany) were therefore exchanged. When compared in homing experiments some differences could be detected at Pisa and Tübingen (Baldaccini and Kiepenheuer, in preparation). The diverging results obtained by Papi and his co-workers in Italy, by Keeton and his co-workers in the USA and by Schmidt-Koenig *et al.* in West Germany suggest that the pigeons' emphasis on different systems or on a different hierarchy of systems, is specific for specific localities and is possibly differently characterized for the pigeons by odours which are stronger in some locations than in others.

There are still doubts about the importance of olfactory cues for the following reasons.

(*a*) The nature of the odorous substances supposedly involved has remained entirely unspecified and it has not been outlined chemically.

(*b*) All evidence of the use of odorous substances is, so far, indirect. Direct evidence, e.g. from laboratory or field conditioning or electrophysiological experiments, that subtle odours such as the ones suggested, can be perceived and differentiated from other odours is lacking.

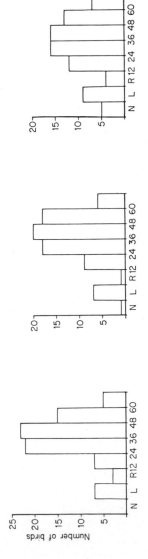

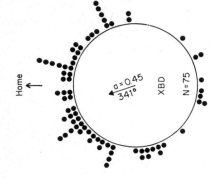

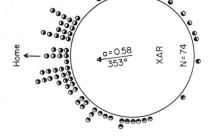

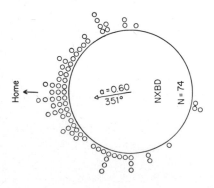

Fig. 67. Summarized initial orientation (circular diagrams) and homing performance (histograms) of eight releases from 40 km. NXBD = no xylocain before displacement; XBD = xylocain before displacement; XAR = xylocain at the release site. The mean vector, its length (a) and sample size (N) are given in the circular diagrams. Homing performance has been summarized in 8 classes of 12 km h⁻¹ and below 12 km h⁻¹ in class. R = homed on the day of release. L = homed after the day of release, N = never home. From Schmidt-Koenig and Phillips (1978).

(c) The representation of olfactory capacities in the peripheral and central nervous system does not appear to be as highly developed as one would have to postulate from the navigational performance.

Such evidence is, however, available for salmon (*Oncorhynchos* spec.) which is known to home olfactorily (e.g. Oshima *et al.*, 1969a,b; Hasler and Wisby, 1951; Idler *et al.*, 1961; Hasler and Scholz, 1978). Of course, further research on birds may well provide some surprises.

The present state of research on olfactory navigation, unsettled as it is, may be summarized as follows. Neither the hypothesis of olfactory navigation in its original form (Papi *et al.*, 1972), nor its later modification, appears to be the universal solution to the navigation problem and it does not take account of all that is known about pigeon homing, as claimed by Papi *et al.* (1972) and Papi (1976). Olfaction is used in homing in pigeons to some extent and Papi is clearly to be credited for this discovery. Olfaction now appears, however, to be only one of several factors in a redundant system, probably a factor of second-order magnitude and possibly in the same order as the use of geomagnetism.

F. Familiar landmarks

A magnetic anomaly, or the characteristic odour of some forest, vineyard or pasture could in some circumstances, be a landmark, but this section refers to visual landmarks only. Though considerable evidence has been accumulated indicating that visual landmarks are not used in pigeon homing (except in the vicinity of the loft), their use is repeatedly suggested, especially when other explanations fail. Matthews (1968, 1974) continues to claim that pigeons pilot by familiar landmarks, at distances at which navigation according to his sun arc hypothesis could not work (for a detailed discussion see Chapter VI, Section B). Papi (e.g. 1976) also reverts to postulating the use of familiar visual landmarks for homing, in those cases where birds still homed after olfactory cues had been excluded or manipulated in one way or another.

Multifold evidence against the use of visual landmarks in pigeon homing has accumulated over the years:

(1) The release site bias of initial orientation persists at most sites, even after repeated releases in very experienced birds (Chapter V, Section H, 1).
(2) Likewise, day-to-day variability continues and is hard to interpret in terms of non-fluctuating landmarks (Chapter V, Sections B, H, 4).
(3) Different homing performances were obtained in simultaneous re-

leases from four directions down to a distance of 5 km from the
loft (Graue and Pratt, 1959).

(4) Clock-shifts predictably deflect pigeons at sites where they had
   been released daily for two weeks (Keeton, 1969), and should
   have developed familiarity with landmarks. Clock-shifts also deflect some experienced birds at distances of 2 km and below, from
   the loft (Fig. 47, Chapter V, A) (Graue, 1963; Schmidt-Koenig,
   1965, 1971; Alexander, 1975). If the loft was directly visible fewer
   birds were deflected, but if the loft was hidden e.g. behind a
   group of trees, more birds followed their shifted compass, although they should have known the trees and other landmarks
   within the range of free flight around the loft from previous
   homing flights.

Results of other experiments seemed to provide somewhat different
evidence. Radio tracking of single birds from the air by Michener and
Walcott (1966, 1967) supported the view that landmarks are not used *en
route*, i.e. the birds fly along different routes after release from the same
sites. Within a few kilometres from the loft, the birds seem to make the
final turn directly toward the loft. The authors associated these turns with
tall buildings in the vicinity of the loft becoming visible (Fig. 68).

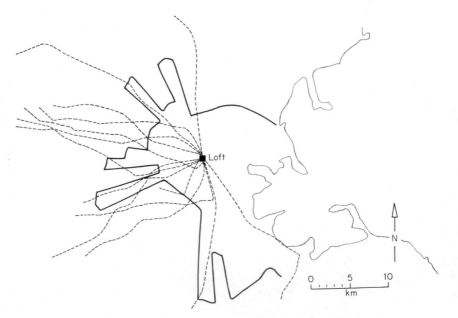

Fig. 68. Final section of tracks (broken lines) of a pigeon near its loft in Cambridge,
Massachusetts. An observer at tree-top level can see the top of the tallest buildings
in Cambridge inside the heavy line. The thin line shows the shore line around
Boston, Massachusetts. From Michener and Wallcott (1967).

Radio-tracking experiments of birds wearing frosted lenses, who were therefore unable to recognize landmarks, showed similar reactions upon reaching the vicinity of the loft. As shown in Fig. 62 and Fig. 69, pigeons react to the vicinity of the loft by either circling, altering course or even perching somewhere after missing the loft. These data suggest that visual landmarks may possibly not be used even in the immediate vicinity of the loft.

Successful homing at night as reported by St. Paul (1962), Keeton (1970b) and Goodloe (1974) also appears to be evidence against the use of visual landmarks.

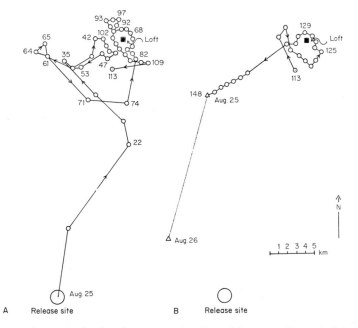

Fig. 69. Approximate track of a pigeon, wearing frosted lenses, radio-tracked from an aeroplane and an additional ground antenna at the loft. The time elapsed after take-off is indicated in minutes at each tracking fix. Tracking fixes were so numerous that they had to be separated into those until minute 113 (A) and those after minute 113 (B). From Schmidt-Koenig and Wallcott (1978).

G. Topography

In contrast to their lack of attention to familiar landmarks, pigeons seem to notice major topographical features such as high mountains, large

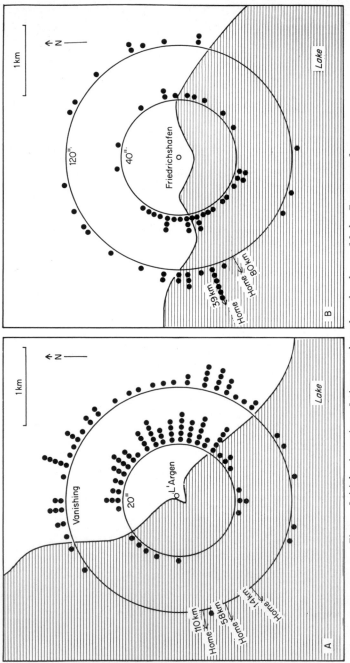

Fig. 70. Initial orientation of pigeons released at the shore of Lake Constance from
A., Langenargen and, B. Friedrichshafen. In A, distribution of birds 2o s after release
and at vanishing are given, in B 40 s and 120 s after release. The 20″ and 40″ circles
are not drawn to scale. From Wagner (1968 and 1972).

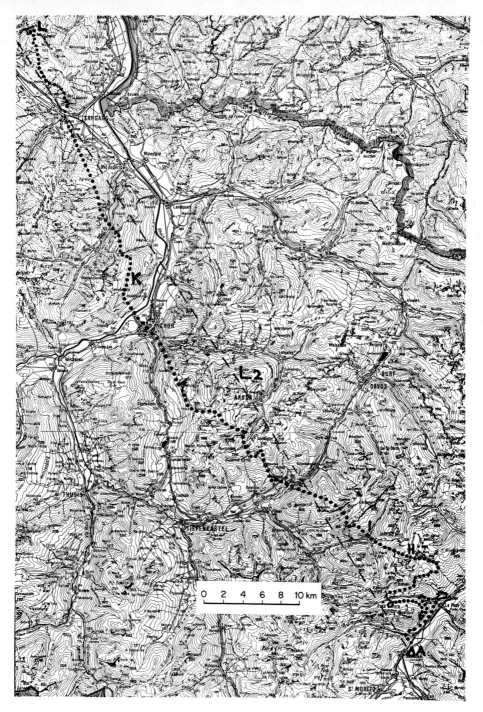

Fig. 71. Track (dotted line) of a group of pigeons obtained by Wagner (1970; 1972) in helicopter tracking in Switzerland. A designates the release site, the dashed line L2, the home direction and L, the location where the birds alighted and tracking was discontinued after 125 km in 93 min. The other letters indicate passes or valleys crossed by the birds. I am indebted to Dr. Wagner for kindly providing the original Figure from Wagner (1970).

Fig. 72. Initial portion of the track (dotted line) of Fig. 71 superimposed on a photograph showing the village of La Punt in the foreground and Piz Kesch towering at the horizon. The pigeons crossed the ridge at Porta d'Escha (H in Fig. 71) to the right (east) of Piz Kesch. I am indebted to Dr. Wagner for kindly providing the original Figure from Wagner (1972).

lakes and the sea coast during the process of initial orientation. In several releases by Wagner (1968, 1972) from the shore of Lake Constance (some summarized in Fig. 70), with home directions pointing across at least 10 km of lake, pigeons generally avoided crossing the water. In other releases in valleys between mountain ridges of the Swiss Alps, the birds followed the axis of the valley, producing vanishing diagrams with one or even two opposite modes. Similar effects have been suggested by others (e.g. Griffin, 1952; Hitchcock, 1952, 1955; Arnould-Taylor and Malewski, 1955).

In extensive visual tracking by helicopter Wagner (1972, 1974) found that topographic structures may also influence the pigeons locally *en route*, but they do not distract them much from the general direction of home. Figure 71 shows the entire track and Fig. 72 the initial portion of the track of a small group of pigeons helicopter-tracked by Wagner (1970, 1972) in the Swiss Alps. Even formidable mountain ridges were crossed at considerable elevations (e.g. Porta d'Escha at 3008 m above sea-level) or valleys at considerable altitudes (800–1000 m above ground) with little deviation from the home direction. Once they seemed to have picked the home direction they also cross bodies of water without much hesitation. When pigeons were released above a layer of fog in the Swiss Alps they did not penetrate the fog downwards, but flew to the nearest peak visible in the distance above the fog (Wagner, 1978). These peaks, however, do not qualify as familiar landmarks.

H. Unexplained phenomena

1. Release site bias

One phenomenon which has never stopped puzzling investigators of pigeon homing is the so-called release site bias. Mean vanishing bearings of initial orientation virtually never coincide with the home direction. Initial orientation and mean vanishing bearings, respectively, are however not independent of the home direction. They do show some relation to home inasmuch as they are, on average, roughly homeward-directed and this may improve with increasing experience (Schmidt-Koenig, 1963d, 1965; Gronau, 1971). Some relationship between homeward directedness and distance of displacement (Chapter V, Section H, 3) has been found, but disagreement between mean bearings and the home directions is usually, despite some possible fluctuations, persistent and characteristic for each site (see also Wallraff, 1959). Soon after the discovery of the release site bias, it was recognized (Kramer, 1957, 1959) that unveiling its secrets may well hold the key to an understanding of what has been called "the map" (Chapter VI, Section F) i.e. the navigational information or basis of pigeon homing.

Keeton (1974a) favoured the view already expressed by Kramer (1959), that release site biases have their cause largely in distorted map information at the release sites, and he tried to investigate possible release site factors. Castor Hill Fire Tower, 143 km north-north-east of the Ithaca loft with its bias 50–90° clockwise from the home direction seemed a particularly suitable site (Keeton, 1973, 1974a). Systematic releases from Castor Hill confirmed and extended previous findings made elsewhere: the bias may decrease somewhat but it does not disappear. The bias persisted under overcast, in birds wearing frosted lenses (Keeton and Schmidt-Koenig, unpublished results) or with bar magnets on their backs. Thus, the release site bias is neither a function of local or general homing experience (although experience contributes to it) nor is it caused by image vision, and the sun and magnetic compasses are not involved (although some biases are associated with magnetic anomalies (Chapter V, Section B). If experienced birds are clock-shifted by the appropriate number of hours clockwise, to deviate from the controls counter-clockwise, just enough to make them depart towards home (Fig. 73B), the homing of the experimental birds was poorer than that of the controls. Eleven control birds and two experimental birds homed on the second day, eight controls and 11 experimental birds homed later. A similar result had been obtained before (Fig. 73A), even at only 22 km distance, by Schmidt-Koenig (1961). The overall pattern including recoveries, indicated that members of both groups flew a clockwise turn not too far out of sight, the experimental birds missing the loft and the controls returning reasonably well. At Castor Hill, normal birds have been radio tracked and found to continue west for 15–20 km to turn southward before reaching Lake Ontario. Such findings indicate that the birds are not making an error within their system, i.e. they read their "map" correctly, but the map is distorted. In extensive clock-shifting experiments by Schmidt-Koenig (1958, 1961) the magnitude of experimental deflections was found to be site-specifically not homogeneous. There were sites at which 6 h clock-shifts typically produced mean deflections of 60° or less, while others showed a 120° mean deflection, where 90° was the theoretical expectation (Fig. 45). Thus. clock-shifts also seem to have their release site bias.

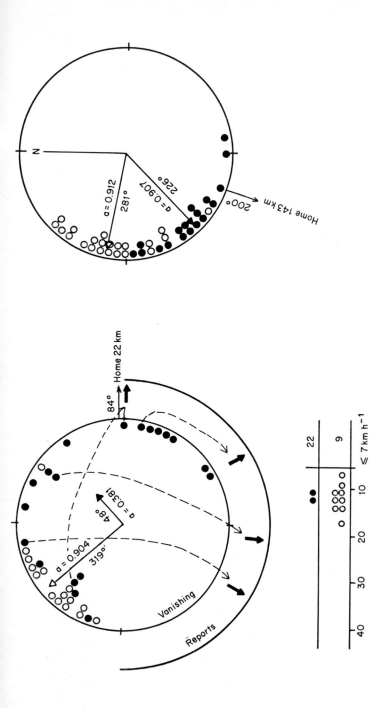

Fig. 73. Release site bias and clock-shifts. Open circles symbolize the vanishing bearings of control birds and solid circles of clock-shifted birds. The mean vectors of each group are indicated.

A. The result of one release from 22 km W. of Osnabrück involving a 6 h shift counter-clockwise. The rectangular diagram shows homing performance. From Schmidt-Koenig (1961).

B. The results of two releases from Castor Hill involving clockwise shifts of 4 h and 5 h respectively. From Keeton (1973).

2. Loft-specific factors

After much descriptive material had been accumulated by Wallraff (e.g. 1959, 1960a) on what was then considered "local effects", the "cross loft experiment" carried out by Schmidt-Koenig (1963c), revealed that some un-known factor at the loft (a "loft-specific factor"), had its share in the re-lease site bias. Data later compiled by Wallraff (1970b) supported this view. Experimental manipulations of the home loft structure resulted in partially predictable biases that may be related to the normal release site biases.

Wallraff (1970a), extending earlier work by Kramer and St. Paul (1954), Kramer (1959), Kramer *et al.* (1959), and Wallraff (1966c) raised pigeons in square aviaries with two parallel walls made of glass or wood, the other two sides either open (mesh wire) or of plastic louvres blocking vision but permitted air flow. The orientation of the two walls (east and west or north and south) produced different initial orientation. The factors involved have not been identified. The deflector loft experiment by Baldaccini *et al.* (1975) discussed in Chapter V, Section E is probably directly relevant here. In Kiepenheuer's (1978a) repetition of this experi-ment at Tübingen, West Germany, the deflection persisted in eight con-secutive releases. As mentioned before (Chapter V, Section E), local anaethesia of the olfactory membrane did not eliminate the deflection, so it can be concluded, at least tentatively, that olfaction is not involved. Homing performance of both deflected groups was as good as that of controls. Thus the manipulation of the loft seems to have established an experimentally distorted map for the pigeon, that worked perfectly well despite being distorted.

3. Distance and direction

Several authors investigated the interrelation of distance and homeward directedness of initial orientation systematically (Schmidt-Koenig, 1966, 1970b; Keeton, 1970a; Graue, 1970), and suitable data from other authors, though much less uniform and extensive, may also be adduced (Wallraff, 1970b; Matthews, 1963c). Windsor (1975) used radio tracking and recorded radio vanishing bearings at 9–13 km distance from the site of release at a time when some corrections of initially false departures had been made by the birds, and scatter was much reduced. His data are, therefore, not strictly comparable to the field-glass vanishing bearings of the other authors.

Figure 74 shows the resulting pattern of field-glass vanishing bearings from seven locations. Despite the unhomogeneous nature of the birds (i.e. experienced and inexperienced) and the experimental designs (i.e. the release was not strictly from the four cardinal directions) used, a fairly characteristic pattern emerges that can hardly be attributed to chance. There seems to be a short-distance zone of good homeward-directed initial orientation, a medium-distance zone of poor homeward-directed initial orientation and homeward-directedness improves again beyond some medium distance, and possibly deteriorates again farther out. Inconsistent with the others and in turn dissimilar to each other, are the curves from Ithaca, New York, and from Bowling Green, Ohio, the latter indicating non-homeward-directed orientation. At the beginning of Keeton's pigeon work at Ithaca, homeward-directed initial orientation *was* found, in contrast to the findings at most other locations. This has evidently changed—at least at some sites—and a new assessment might turn out results more

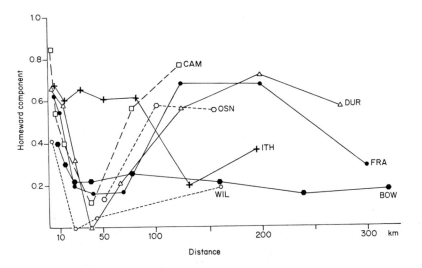

Fig. 74. Homeward directedness of initial orientation (ordinate) as a function of distance (abscissa) at seven locations. BOW = Bowling Green, Ohio (Graue, 1970); CAM = Cambridge, England (from Matthews, 1963c); DUR = Durham, North Carolina (from Schmidt-Koenig, 1966); FRA = Frankfurt/Main, W. Germany (from Schmidt-Koenig, 1970b); ITH = Ithaca, New York (from Keeton, 1970a); OSN = Osnabrück and WIL = Wilhelmshaven, W. Germany (from Wallraff, 1970a). Homeward component h = +1·0 would mean that all birds headed precisely towards home; h = −1·0 that all birds headed precisely away from home; h = 0 that the birds either headed ±90° of the home direction or else randomly. Solid lines connect mean values of h, for releases from the four cardinal directions in systematic experiments with experienced birds; broken lines connect mean values of h for only 2 releases, from roughly opposite directions, with inexperienced birds (with the exception of CAM, which also involved experienced pigeons). From Wallraff (1974a).

in agreement with the others. The cause for that change is unclear. Experimenting may have selected for or against some modes of orientation, or some unknown change at the release site, or of the cues available, may have occurred in the course of time (Keeton, pers. comm.).

Keeton (1974a) is correct in arguing that the homeward component is a function of two quantities, the deviation of the mean direction from the home direction and the degree of scatter of the individual bearing. Nevertheless, the homeward component is a measure of homeward directedness and there is also the overall interrelation with distance as shown in Fig. 74. Wallraff (1967) and Keeton (1970a) point out that the curves summarizing data from all four directions do not reveal what may be direction-specific differences of orientation towards home. Indeed, as discussed by Schmidt-Koenig (1970b), despite the fairly good agreement between the DUR and the FRA overall curve (as shown in Fig. 74) there

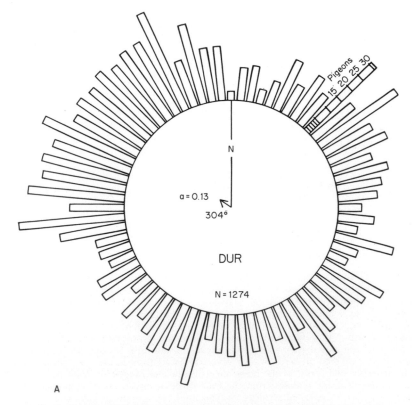

Fig. 75. Summarized vanishing bearings from 64 symmetrical releases of experienced birds domiciled, A, at Durham, New York (DUR) and, B, at Frankfurt, W. Germany (FRA). For more details see text. From Schmidt-Koenig (1970b) and unpublished.

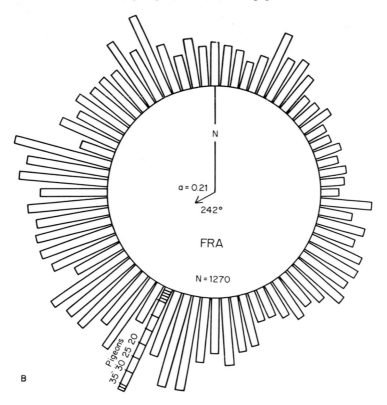

are differences in detail. Summaries of data compiled differently to Fig. 74 reveal significant overall differences between DUR and FRA. All vanishing points of the DUR and the FRA series of Schmidt-Koenig are summarized to geographic north in Fig. 75. Both are non-random ($P < 0.01$) with mean directions pointing in westerly directions, west-south-west for FRA and north-west for DUR. Both diagrams are significantly different from each other ($P < 0.01$). A westerly bias is superimposed on initial orientation, though it is significantly different for the two locations. All the more striking is the fairly good overall agreement of the distance-dependent pattern of orientation towards home. Its basis remains unknown, but a bias superimposed on initial orientation seems to be a general phenomenon contributing, to some extent, to the release site bias and masking the distance effect. Whether or not their basis is loft-specific or a characteristic of the general area is unknown. The pattern as evident from Fig. 74 may indicate the operation of two navigation systems, one for shorter and the other for longer distances. In the intermediate zones neither works satisfactorily.

4. Temporal fluctuations

In addition to the variations attributed to aperiodic variations of the geomagnetic field (Chapter V, Section B), or periodic variations associated
with the lunar day (Chapter V, Section C), unexplained aperiodic variations of several kinds have been described. Wallraff (1959) observed variations in terms of years, days and hours involving initial orientation as
well as homing performance. Attempts to correlate these fluctuations with
several environmental variables were unsuccessful. More recently Wallraff
(1971) found seemingly aperiodic variations of initial orientation of less
than 10 min. Sequences of two or more directionally similar bearings of
pigeons released at 5 min intervals were more frequent than can be expected by chance. The nature of the assumed environmental factor and
the mechanism of interaction with pigeon orientation remained unknown.
From the periodic fluctuations, annual cycles of homing performance are
well-known to all pigeon racers, and they have been confirmed experimentally by Kramer and St. Paul (1956) and Wallraff (1959). Gronau and
Schmidt-Koenig (1970) and Gronau (1971) reported that the homeward
directedness of initial orientation revolves annually, like homing performance (Fig. 76). In these experiments many variables that also influence
homing performances have been kept constant by using only one release
site and groups with well-defined states of experience. The data were
accumulated over almost three years and the cycles were more pronounced in naïve birds than in experienced homers. The annual cycles
roughly paralleled that of ambient temperature being best in summer,
even in hot summer climates, such as in North Carolina (Schmidt-Koenig,
unpublished results) and poorest in winter, but a continuous correlation
between ambient temperature and performance does not seem to exist.

Closing the discussion on observed fluctuations, underlying causes have
been found only for some small fluctuations: the natural variability of
the geomagnetic field and the lunar cycle possibly related to gravitation,
is involved in some way in pigeon homing.

I. Direction of the preceding release

Learning has its share in pigeon homing: experienced pigeons perform
better in every respect than inexperienced birds. Each release adds experience until a certain level of asymptotic further improvement only is

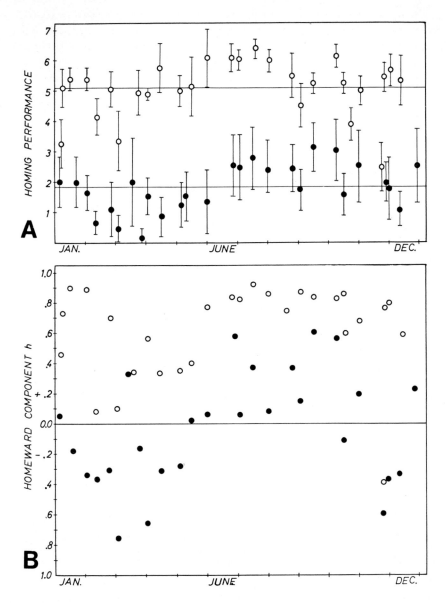

Fig. 76. A. Mean relative homing performance (ordinate in class units) of experienced
(○) and inexperienced (●) birds. The 99% confidence interval is given at each
symbol of n = 20. The lines show the overall mean of veterans (upper line) and of
inexperienced birds (lower line). Relative homing performance: o would mean that
no bird homed, 7 that all birds homed at >60 km h⁻¹. B. Homeward component h
(ordinate) as a function of season (abscissa) of the same releases as in A. Confidence
intervals cannot be calculated. The homeward component h is explained in the
legend of Fig. 74. From Gronau and Schmidt-Koenig (1970).

reached. Several authors reported that pigeons when released repeatedly from the same "training" direction, continue to fly in that direction when subsequently released from sites off the training direction (e.g. Kramer and St. Paul, 1950b; Matthews, 1951, 1955; Hitchcock, 1952; Riper and Kalmbach, 1952; Pratt and Thouless, 1955; Michener and Walcott, 1966, 1967; Benvenuti *et al.*, 1973a). Pratt and Thouless (1955) and Wallraff (1959) suggested that even a single homing flight may produce a measurable bias on the initial orientation of the next release. Wallraff (1959, 1967, 1974b) had only incidental data of questionable quality (for a detailed discussion see Schmidt-Koenig, 1965 and Alexander and Keeton, 1972) while Graue (1965), Alexander and Keeton (1972) and Schmidt-

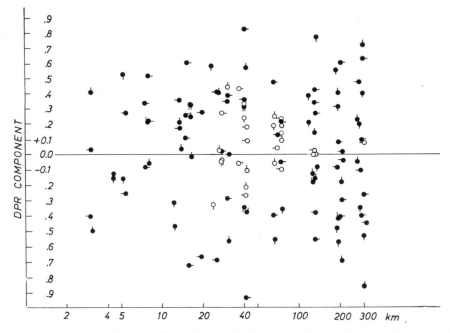

Fig. 77. Direction of the preceding release on initial orientation of the next release. The component of the direction of the preceding release (ordinate) as a function of the distance of release (abscissa, logarithmic scale) of 128 releases of experienced pigeons. Bristles at each score indicate the direction of release (e.g. from the north). White scores identify releases that were random (1% level). The component of the home direction of the previous release (DPR component) for each vanishing diagram of 20 (19) birds was calculated as $C_{DPR} = a \cos (\alpha - \phi)$ where a represents the length and α the direction of the mean vector of each vanishing diagram and ϕ represents the home direction of the preceding release. If all birds headed precisely into the home direction of the preceding release, (i.e. 90° left or right of the actual home direction) C_{DPR} would be $+1.0$. Values $+1.0 \geqslant C_{DPR} > 0$ represent various degrees of mean headings in the semicircle of the home direction of the preceding release. Values $0 \geqslant C_{DPR} > -1.0$ would indicate mean headings in the semicircle opposite of the home direction of the preceding release. From Schmidt-Koenig (1976).

Koenig (1976) contributed data from a series of releases designed to answer this question. Graue's well-designed experiments qualitatively supported the notion that each preceding release had an effect on the next, inasmuch as the bias observed was always in the expected direction. The deviations were, however, numerically small and frequently insignificant. Alexander and Keeton (1972) found no significant effect and so did Schmidt-Koenig (1976), who provide the largest batch of data. In Fig. 77 the DPR component (component of the direction of the preceding release; calculation explained in the legend of Fig. 77) is plotted as a function of the distance of release (as a variable relevant in Fig. 74). If the home direction of the preceding release had an effect on the initial orientation of the following release, a significant majority of scores should be larger than zero, or the arithmetic mean of all scores should be significantly larger than zero. There is no trace of such an effect. Alexander and Keeton and Schmidt-Koenig agree that the direction of the preceding release can be considered not to contribute much, if at all, to the bias of initial orientation which is frequently observed.

J. When and where does the pigeon navigate?

The questions whether pigeons collect navigational information during the outward journey, at the release site before release, or whether they perform navigation upon release, or *en route* home, have been asked repeatedly. Inertial navigation would require collecting information during the outward journey; if pigeons were using the sun arc navigation system, they could do so only upon release, and so on. An answer to the question raised above might exclude a number of alternative hypotheses and help us to concentrate on a few only. It now appears that navigational processes may take place at all stages. Magnetic and olfactory information may be sampled during the outward journey (Chapter V, Sections B, 2 and E). In an unpublished analysis of data from Schmidt-Koenig (1970b) the last five pigeons homed significantly faster than the first five of a sample of 20 released within 2–3 h, possibly indicating that navigational information may be perceived or processed by the birds sitting in canvas-covered crates at the release site. Keeton (1974a) routinely subjects large numbers of releasing and homing data to correlation analyses. He found an inverse correlation between time of release and vanishing time, supporting the view that birds may obtain or process navigational information while waiting at the release site. Correlations between vanishing bearings and homing speeds as reported by Wallraff (1959) could not be confirmed.

E*

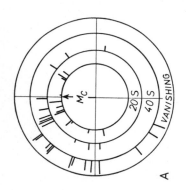

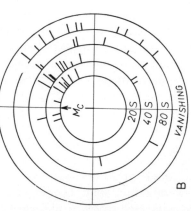

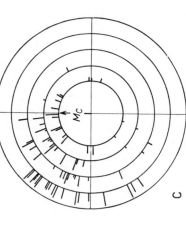

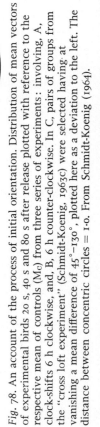

Fig. 78. An account of the process of initial orientation. Distribution of mean vectors of experimental birds 20 s, 40 s and 80 s after release plotted with reference to the respective mean of controls (M_C) from three series of experiments: involving, A, clock-shifts 6 h clockwise, and, B, 6 h counter-clockwise. In C, pairs of groups from the "cross loft experiment" (Schmidt-Koenig, 1965c) were selected having at vanishing a mean difference of 45°–130°, plotted here as a deviation to the left. The distance between concentric circles = 1·0. From Schmidt-Koenig (1964).

Bearings of initial orientation intermediate between take-off and vanishing, at the 20 s, 40 s and 80 s intervals after release, were recorded by Schmidt-Koenig (1964) and are plotted in Fig. 78. Even 20 s after release, control and experimental birds had separated, and the vanishing direction is attained, on average, 80 s after release. This may simply reflect the time requirement for a compass orientational performance but it is likely that it also involves navigational processes. Finally, corrections of initially false directions must take place *en route* as generally indicated by successful homing of birds that did not, and ordinarily do not, depart precisely toward home. Some more detailed aspects of corrections *en route* have been discussed in Chapter V, Section H, 1.

Thus, navigational information seems to be processed in all stages and answers to the original question are not very helpful in eliminating or segregating certain modes of navigation. However, the process of initial orientation is as yet little understood. It is inadequately represented by

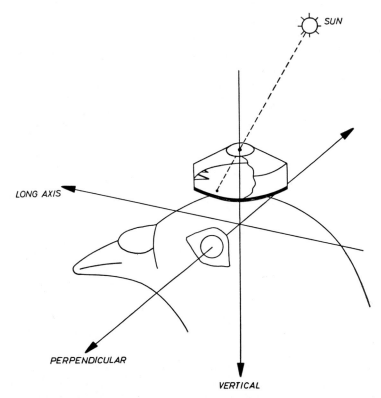

Fig. 79. Camera attached to the pigeon's head to record direction of fixation during initial orientation as developed by Köhler (1978). A shutter closed the aperature after 60 s.

vanishing bearings which are routinely recorded as "snap shots" by field-glass observation at the limits of human visual power. Even if intermediate bearings at some intervals such as 20 s, 40 s and 80s after release are recorded, the account remains incomplete and a lot of possibly valuable information is wasted. Elsner (1978) developed a visual tracking system using two stations with sights for triangulation, and electronic systems for automatic processing and plotting the track of a bird at intervals of less than 1 s. Only a few preliminary data have been collected so far.

In order to study the head movements of pigeons upon release Köhler (1978) developed an ingeniously simple camera to be attached onto the bird's head (Fig. 79). Through the aperture on top of the camera the sun forms a spot on the horizontal film each time the bird holds its head steady for a few seconds. With the position of the sun known from astronomical tables, the direction of fixation can be calculated. Preliminary data presented by Köhler (1978) show that the pigeons keep their heads steady for short periods of time more frequently in clear weather than in hazy conditions. Köhler assumed the angle of fixation in pigeons to be 90° (it is actually nearer 75°). It is, therefore, not yet clear whether the direction of fixation can be related to distant landmarks or to some other (possibly non-visual) feature relevant for initial orientation. Köhler's results tentatively indicate that some processes important for navigation take place immediately after release.

VI. Hypotheses, theories, and concepts

A. Coriolis force and the geomagnetic field

With the growing evidence of the role of the geomagnetic field in compass orientation and navigation, Yeagley's (1947, 1951) hypothesis needs to be discussed though in all probability more for historical reasons than for an actual role in bird navigation.

The geophysical basis of this theory is as follows: the lines of equal strength of the vertical component of the geomagnetic field are roughly circular and centred on the magnetic poles. The lines of equal Coriolis force are similarly centred on the geographic poles. Since the geographic and magnetic poles are approximately 2300 km apart from each other, the two sets of lines form an irregular grid (Fig. 80) that could, theoretically, be used for navigation. This grid is similar to the grid of geographical latitude and longitude, though in the former the lines intersect at angles far from rectangular and they intersect more than once.

Yeagley (1947, 1951) suggested that the birds navigate by measuring the local strengths of both variables. In one type of experiment to test this hypothesis, Yeagley attached magnets to the wings of experimental birds to jam magnetic detection, but in only one of two experiments was there any effect. Another type of experiment made use of the fact that there were so-called conjugate points at which vertical magnetic and Coriolis isolines intersected twice, resulting in locations with very similar magnetic and Coriolis values, and consequently indistinguishable for the birds. Yeagley released birds from a loft in Pennsylvania near such a conjugate point in Nebraska. Reanalysis of this data showed that, contrary to his claim, birds were not orientated towards that conjugate point (Kramer, 1948; Griffin, 1952; Matthews, 1951, 1955). Whether or not the Coriolis force could be measured by a flying bird has also been a point of considerable criticism on theoretical grounds (e.g. Davis, 1948;

Slapian, 1948; Varian, 1948; de Vries, 1948; Griffin, 1952), after Ising (1945) and Thorpe and Wilkinson (1946) had put forward theories and calculations as to how the Coriolis force might be utilized for navigation.

Yeagley's hypothesis has been discarded, and the entire discussion discredited the use of geomagnetism for orientation for more than a decade.

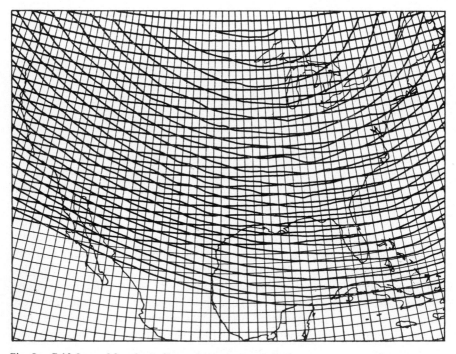

Fig. 80. Grid formed by the isolines of magnetic vertical (stronger curved lines) and of Coriolis forces for North America. From Yeagley (1947).

B. Sun navigation

Matthews (1953b, 1955) advanced the sun arc navigation hypothesis, which derives its name from the suggestion that pigeons extrapolate a short portion of the sun's arc, measured at the displaced position, to the noon altitude (Fig. 81) and compare it with the altitude last seen at home for latitudinal displacement. If the noon altitude observed is lower than that remembered from home (as sketched in Fig. 81) the bird would be north of home. If the noon altitude observed is higher than that of home

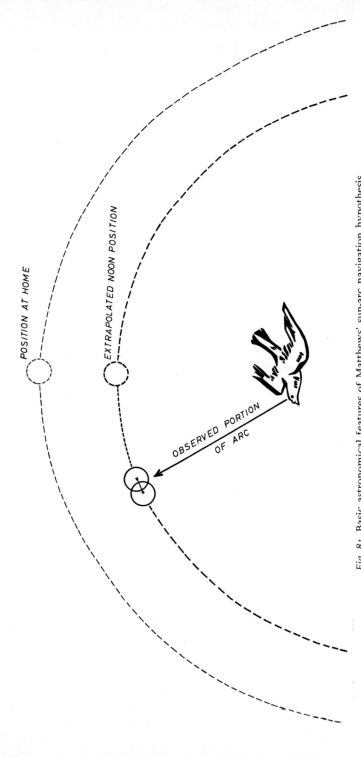

POSITION AT HOME

EXTRAPOLATED NOON POSITION

OBSERVED PORTION OF ARC

Fig. 81. Basic astronomical features of Matthews' sun-arc navigation hypothesis.

the bird would be south of home. Similarly, an angular or time difference between actual azimuth position and extrapolated noon azimuth of the sun, as compared to that at home, would indicate a displacement east or west of home.

As support for his hypothesis, Matthews designed and carried out firstly, several clock-shift experiments and secondly, two sun occlusion experiments.

Clock-shifts clockwise (or counter-clockwise) were supposed to simulate a displacement east (or west). Birds should, consequently, head west (or east) upon release. Recalculations of Matthews' original clock-shift experiments reveal that the data do not support this conclusion. In only one out of five diagrams obtained by Matthews (1955), did the experimental birds perform non-randomly at the 99% level (Schmidt-Koenig, 1965). Meanwhile clock-shifting experiments have been repeated many times by several authors (e.g. Schmidt-Koenig; Keeton; Walcott) and with many modifications in terms of degrees of shift, conditions of shifting (e.g. Alexander and Keeton, 1974) and distance and direction of release. All have shown that the sun is used for compass orientation and not for navigation (Chapter V, Section A).

The sun occlusion experiment was based on the rather arbitrary assumption that pigeons remember only the noon altitude that they saw last at home. In this experiment a group of pigeons was prevented ("occluded") from seeing the sun for a number of days at the autumn equinox. At this time of the year the noon altitude of the sun as a basis for latitudinal reference decreases rapidly. During the period of occlusion the noon altitude had fallen sufficiently low so as to simulate a northward displacement, even if the birds were actually taken some distance to the south. The occluded birds should, therefore, fly south, i.e. away from home. The majority of Matthews's (1953b) experimental birds did fly southward, i.e. away from home, in one experiment without controls, and in another experiment, all but four experimental birds flew in the opposite direction to the controls (i.e. also away from home). Hardly mentioned, however, is the fact that the experimental birds homed as fast as the controls. Thus, even this experimental evidence directly supporting Matthews's hypothesis rests on rather weak grounds. Moreover, half a dozen repetitions of the autumn equinox sun occlusion experiment (Rawson and Rawson, 1955; Kramer, 1955, 1957; Hoffman, 1958; Keeton, 1970c) and three additional spring equinox experiments (Keeton, 1974a), yielded negative results (Fig. 82). Matthews (1974) qualifies this evidence as "unclear and conflicting".

Walcott and Michener (1971) used a different approach to examine the effect of an incorrect sun altitude seen at home before displacement. They confined single pigeons in a cage from which they could see a mirror

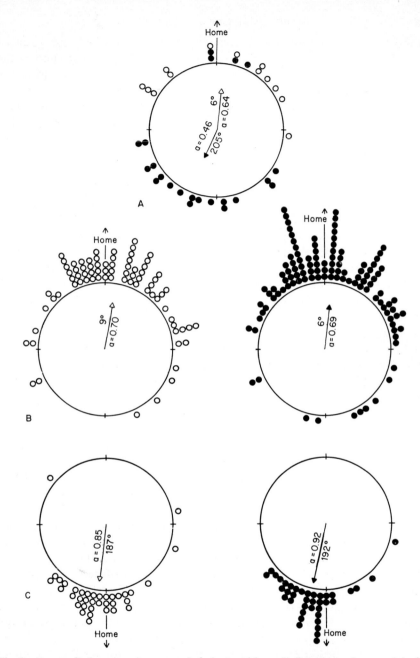

Fig. 82. Sun occlusion experiments and their repetitions. A. Summary of two original autumn equinox experiments by Matthews (1953b). B. Summarized autumn equinox repetitions by Rawson and Rawson (1955), Kramer (1955, 1957), Hoffmann (1958) and Keeton (1970c). C. Summarized spring equinox experiment by Keeton (1974a). Open symbols are used for control birds, solid symbols for experimental birds. For more details see text.

view of the sun around noon. The mirrors were attached and adjusted in such a way that the sun appeared 31–70' higher (or lower) than the local sun at noon. This altered noon altitude corresponded to locations 58–130 km south (or north) of the true loft position. The birds were exposed to this shift of noon altitude for 10–14 days. When released from sites between the true loft position and the simulated home position experimental birds headed for home, again contrary to what would be predicted for sun arc navigators.

Whiten (1972) trained pigeons in key-pecking conditioning experiments in a rotating apparatus, to relate sun altitude to the home direction from sites north and south of home. The results may be interpreted as evidence for a navigational performance, but they may equally be interpreted as showing that pigeons can be trained to respond to differences in sun altitudes, i.e. to discriminate between a low and a high altitude—which has been noticed before—and to associate this discrimination with a north–south discrimination. Additional experiments will have to decide between these alternatives. It would be premature to adduce Whiten's results as support for sun navigation, and it would be the only support to date against overwhelming negative evidence.

Other aspects not in agreement or directly against Matthews's hypothesis have been extensively and repeatedly discussed in the literature (e.g. Kramer, 1957; Pennycuick, 1960; Schmidt-Koenig, 1965, 1970c, 1975; Keeton, 1974a), prompted by Matthews' (e.g. 1968, 1971, 1974) untiring attempts to propagate his hypothesis and to discredit unfavourable experimental evidence or discussion by others.

The major arguments against Matthews' hypothesis may be summarized as follows. Firstly, unlike Pennycuick's (1960) modification, Matthews' sun arc hypothesis would require measuring changes in azimuth. There is no evidence that a flying bird can do this.

Secondly, there is likewise no evidence that birds possess a rigidly stable chronometer maintaining home time, or a time equivalent to Greenwich mean time, in human astronomical navigation. Countless clock-shift or similar experiments demonstrated that the sun compass clock is anything but resistant to phase shifts. If the animal is displaced east or west, that clock is synchronized with the new local time in a matter of days. Many circadian biorhythms—the sun compass clock is just one of them—have been investigated under constant light. They all turned out to drift out of phase with local time, i.e. they are anything but stable.

Thirdly, the high degree of accuracy necessary for keeping record of the time of day and precisely measuring time intervals of minutes and seconds, which is required for sun navigation, has not been found.

Fourthly, extraordinary acuity of vision would be essential for sun navigation. Apart from laboratory and anatomical findings to the contrary

(Chapter III, Section A, 5), homing experiments with frosted lenses cast considerable doubt on the claim that the pigeon's eye is adequately endowed and that precise vision is involved in homing at all. Recent reconsiderations of a possible role of the pecten oculi in orientation by Pettigrew (1978), including a model of sextant function of the pecten, are discussed in Chapter III, Section B, I. They do not alter the balance of arguments against sun navigation.

I would like to emphasize and acknowledge the high academic merits of Matthews' hypothesis. It stimulated much experimental and theoretical work between the early fifties and the early seventies. But it is not the answer to the question of how pigeons, or birds in general, navigate.

As mentioned earlier, one stumbling block in Matthews' concept was the need for measuring changes of sun azimuth in flight, to obtain some portion of the sun's arc for extrapolation to the noon altitude. To get round this practical impossibility Pennycuick (1960) advanced a modified version of the sun navigation hypothesis, suggesting that the displaced bird measures firstly, the actual sun altitude to obtain latitudinal information and secondly, the rate of change of altitude for longitudinal information, both to be compared with the corresponding quantities at home. This variation avoids the necessity for measuring changes of azimuth. It is, on the other hand, burdened with the same problems as Matthews' hypothesis, in that the bird must be able to measure minute sun movements and time intervals and maintain home time all with very high precision, and have an almanac-like memory of relevant sun data. Pennycuick suggested specific experiments to test the hypothesis, but none have ever been carried out. In any event, the same experimental evidence that stands against Matthews' hypothesis also stands against Pennycuick's variation.

C. Star navigation

Kramer (1949) discovered that the migratory restlessness of caged nocturnal migrants is directional rather than random. Sauer and Sauer (1955) (also Sauer, 1956, 1957) later confirmed this under the natural sky and under artificial planetarium skies, using European warblers of the genus *Sylvia*. Others confirmed this under the natural sky for other species (Perdeck, 1957; Mewaldt and Rose, 1960; Hamilton, 1962b, 1966; Sauer, 1963; Shumakov, 1965, 1967; Emlen, 1967a,b,c, 1972, 1975; Sokolov, 1970; Rabol, 1969, 1970, 1972; Wiltschko and Wiltschko, 1975a,b). All demonstrate the birds' ability to maintain roughly the migratory direction appropriate for the season, i.e. southerly in autumn and northerly in spring.

On testing warblers under planetarium skies which were manipulated to simulate some east or west displacements and to simulate parts of the autumn southward migration, Sauer (1957, 1961) and Sauer and Sauer (1960) proposed a complete bico-ordinate star navigation system utilizing either azimuth and altitude, or hour angle and declination of stellar patterns. However, theoretical considerations and experimental data do not support Sauer's hypothesis. A stable navigational clock, precise timing of short intervals, and high acuity of vision would be required for this hypothesis but have not been demonstrated. Sauer considered as principal support for his hypothesis, autumn planetarium experiments in which he exposed one lesser whitethroat (*Sylvia curruca*) to skies of decreasing latitude so as to simulate a southward displacement. The bird indicated initially southeasterly and, at lower altitudes, southerly directions. Even when taken at face value, recalculation by Wallraff (1960b,c), revealed that Sauer's geographical interpretation of a migratory detour east around the Mediterranean accomplished by astronavigation, was unwarranted. The data were in agreement with a more or less direct route and the turn to more southerly directions at lower latitudes was neither as far north nor as large as Sauer claimed (it was not significant). Sauer's method of direct observation, the lack of a clear criterion of what to consider an oriented reaction, the improvised recording, the lack of statistical independence of the raw data, all precluded statistical treatment. This and the small samples by no means support the far-reaching conclusions drawn by Sauer. Sauer's (1963) subsequent experiments with golden plovers (*Pluvialis dominica fulva*) did not turn the tide despite improved recording techniques.

Meanwhile, extensive studies by various investigators (e.g. Emlen, Wallraff, Wiltschko), using different methodological approaches and different species of birds (discussed in Chapter III, Section B, 3), have made it abundantly clear that stars are used for directional or compass orientation and not for navigation.

D. Inertial navigation

It is an old idea—it was considered as early as 1873 by Charles Darwin—that an animal could record all turns and distances, either while passively being displaced, or during some actively-performed journey, and could precisely retrace all movements (the principle of Ariadne's thread), or integrate direct direction and distance from home. For the second alternative Barlow (1964, 1966) has spelled out a detailed hypothesis. The basic principle is to record distance and direction travelled through double integration of acceleration over time. The vertebrate vestibular apparatus is capable of measuring accelerations but the necessary sensi-tivities for pin-point navigation have not been demonstrated and the central integrating mechanisms for processing the information from the vestibular apparatus remain hypothetical to date. These considerations do not warrant a final verdict. First of all, there may be systems other than the vestibular apparatus sensing acceleration. Secondly, the sensitivity may be sufficiently good, but this has not been demonstrated. The use of the geomagnetic field for orientation had been rejected for many years on similar grounds of lack of sufficient sensitivity. Meanwhile, we know that biomagnetic reception reacts to changes of the order of considerably less than 0·001 Oe (Chapter III, Section B, 4).

The experimental evidence appears to be clearly negative. Several authors have rotated birds on a turntable during displacement (Rüppell, 1936; Griffin, 1940; some more unpublished experiments are mentioned in Keeton, 1974a), assuming that rotation would introduce an error into the birds' inertial system, which would be expressed in poorer orientation of experimental birds than that of controls, but no difference was reported.

The same intention to introduce an error into or to "exhaust" the iner-tial system, made several authors displace birds by way of large detours or by aeroplane in strictly clockwise or counter-clockwise spirals (Schmidt-Koenig, unpublished data). A fairly consistent effect of detour experiments has been reported by Papi *et al.* (1973), Fiaschi and Wagner (1976) and Papi (1976) in experiments attributed to olfactory homing. Others found either no effect, such as Keeton (1974c) or inconsistent effects (Hartwick *et al.*, 1978; Keeton and Papi, pers. comm.).

Experimental interference with the supposed inertial input, such as displacement under full anaesthesia, has not produced any difference between experimental and control birds (Exner, 1893; Kluijver, 1935; Griffin, 1943; Wallcott and Schmidt-Koenig, 1973) although this treat-ment should also have interfered with the proposed olfactory input during displacement.

The most direct experimental action on the vestibular apparatus, by surgically bisecting semicircular canals or removing the cochlea, lagena or sacculus, yielded consistently negative results (Hachet-Souplet, 1911; Sobol, 1930; Huizinger, 1935; Wallraff, 1965, 1972b). To exclude the possible alternative use of solar data, Money and Keeton (in Keeton, 1974a) performed experiments, under overcast, with birds after the removal of the sacculi. No difference between experimental and control birds was found (see Fig. 83).

The evidence in this area is not entirely convincing: in all the experi-ments above, some loophole remained for alternative explanations. The bird's system for detecting and integrating accelerations may be suffi-ciently sensitive and accurate not to be upset by circuitous detours or

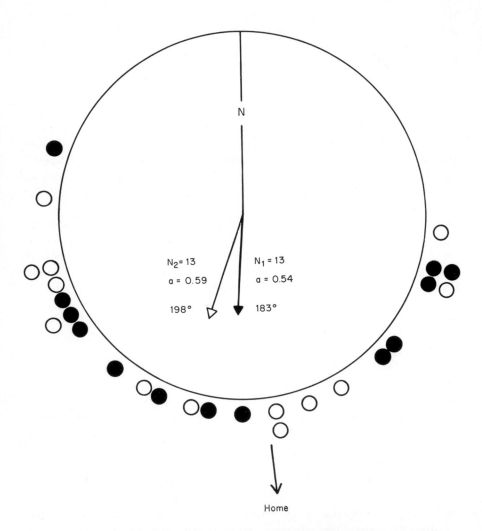

Fig. 83. Initial orientation of experimental pigeons (solid symbols) after removal of their sacculi, and control birds, under overcast skies, in an experiment on the possible use of inertial forces during the outward journey. From Money and Keeton, in Keeton (1974a).

rotation on turntables. Even full anaesthesia may not be sufficiently deep to eliminate recording of acceleration, and cutting one semi-circular canal or removal of some other part of the labyrinth still leaves the possibility that acceleration is measured with the remaining canals, or that the sensors of acceleration are located entirely elsewhere. Though the evidence is clearly negative and most experts agree that inertial navigation has a very low probability of actually existing, they also agree that the final convincing experiment to disprove this hypothesis has yet to be carried out.

Before leaving this topic I should mention an experiment that may be relevant here. Gualtierotti *et al.* (1959) and Schreiber *et al.* 1962) examined the cerebellar response of birds to rotation. In homing pigeons and migratory species (but not in sedentary species), prolonged discharges were noticed after rotation on a turntable. While not meaning very much *per se*, these findings may be related to perhaps simpler forms of usage of inertial information, as discussed by Keeton (1974a). There is good evidence that birds continue to fly along straight lines even inside clouds or other conditions excluding visual or environmental help. Inertial information of a lower quality than that required for navigation may be the alternative solution here.

E. Olfactory navigation

The hypothesis advanced and later modified by Papi *et al.* (1972, 1973) seems to be very simple. Pigeons and possibly other birds as well, create an olfactory map in which each area is composed of a characteristic pattern of volatile, presumably organic, odorant substances. In an early phase of their lives they learn to associate odours with the directions the odours come from. During displacement they keep track of the direction of displacement by recording the various odours through which they are being transported, and they reverse and integrate the directions to obtain the home direction. The implications of the hypothesis are not as simple as they seem to be at first glance. All the experiments which are discussed in Chapter V, Section E are merely indirect proofs of the birds' supposed olfactory capacities. There is no direct evidence that birds can discriminate odorous substances originating from, say, a forest 3km north of the loft from those of a forest 2 km east of the loft, or the odour from a pasture from that of a suburb in a mixture of different and subtle odours.

Further critical remarks have been made in Chapter III, Section A, 3 and Chapter V, Section E.

F. Map and compass

The concept of map and compass navigation was proposed by Kramer (1953). After having discovered the sun compass in birds and realizing that a compass by itself would hardly be sufficient for navigation, he visualized a bird to be in a similar position as a boy scout knowing his location on a map, aligning the map and heading for home with his compass. "Map" was a term used to mean the navigational information telling the bird where, in relation to its goal, its present position was. Kramer left the properties of the map and even the sensory modalities involved on the part of the bird entirely unspecified. Several investigators considered the map literally as some imaginary unit (e.g. Wallraff, 1959, 1960a). Later, with the advance of theoretical and experimental analysis, the map was resolved into information on direction and distance Schmidt-Koenig, 1963c,d, 1966, 1970b; also Barlow, 1964) or, possibly, geophysical variables providing a grid of co-ordinates.

Wallraff (1974a) made an attempt at a systems analysis of the "map", assuming that navigation is based on position-dependent bico-ordinate information and not on information collected during the outward journey. According to his null-axis concept, the map may be visualized as a bico-ordinate grid with the null-axis co-ordinate going through the goal but not necessarily at the same time through the displaced position. The bird would first attempt to reach the null-axis and then follow it until intersection with the second co-ordinate signals that the home position has been reached. This is only a theoretical modification of earlier suggestions of bico-ordinate navigation systems. It is sufficiently unspecific to cover many of the puzzling phenomena of variability observed in pigeon homing. The physical basis and sensory modalities involved remain open. Direct experimental support for some sort of a null-axis is not available, but there is increasing evidence that navigational information is collected during the outward journey and not exclusively at the release site (Chapter V, Section K).

G. Vector navigation

A vector is a quantity characterized by direction and length. If, at some starting point, one is informed on the direction and possibly on the distance to travel, one can get to a goal at the tip of that vector (Fig. 84). If *en route* from S to G in Figure 84 some displacement, e.g. by a storm or by some experimenter takes place, G¹ is reached instead of G. Thus vector navigation is a system of only limited scope and it is a matter of definition whether or not it deserves the term navigation. Experimental work as discussed in Chapter III, Section B, 6 and Chapter IV, Section B does, however, support the existence of vector navigation. If the abstract graph of Fig. 84 is superimposed on Fig. 42 it would coincide with Perdeck's findings in starlings and support the interpretation that juvenile birds reach their winter range by vector navigation. The information on the distance may be provided by information on the quantity of migratory restlessness or time spent in migration as discussed in Chapter III, Section B, 6. Thus, there is some support for the idea that vector navigation does operate and it seems to offer a partial solution to the problem of bird migration, i.e. that partial aspect that concerns the first autumn migration of juvenile birds.

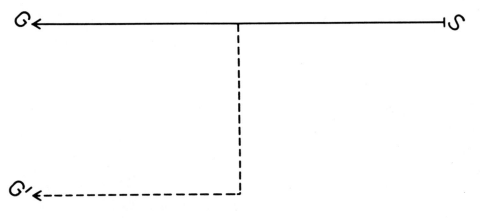

Fig. 84. Diagrammatic representation of vector navigation. S = starting point; G = goal; G¹ = destination upon intermittent displacement (perpendicular dashed line).

VII. Conclusion

Directional orientation plays an important role in migration as revealed by radar, by laboratory studies and also by pigeon homing. The mechanisms used in directional orientation are three compasses: the sun compass, the star compass and the magnetic compass. They are alternative mechanisms utilized by the birds, depending upon the availability of the appropriate cue, in a hierarchical order. All three are established ontogenetically in different ways and operate on different principles (visual, non-visual, time-compensated, not time-compensated etc.). Their existence suffices to explain directional orientation in birds. Vector navigation, which integrates compass orientation and information on the distance to travel, in terms of amount of activity or time spent in migration, is of limited scope. It may explain some aspects of bird migration, for example the first migratory journey to the winter range of juvenile birds. Vector navigation is insufficient to enable birds to carry out true navigation.

Several environmental variables have been shown to be utilized for true navigation or homing: geomagnetic cues, olfactory cues, and variables in the atmosphere, which are deflected or reflected by wooden walls or other materials impermeable to wind. It may well be that the cues known at present suffice for navigation (or at least for the special case of homing in pigeons), but we cannot yet appreciate how the bird utilizes and integrates these variables into its navigation system. It may also be the case that all the variables segregated or suggested so far are factors of second-order magnitude only and the first-order factors are still unknown. Whichever of these alternatives is in fact correct, we are as yet unable to explain navigation and homing in birds.

VIII.   Summary

I. Introduction to migration

Birds migrate in the southern as well as in the northern hemisphere. Migratory routes only exceptionally run precisely on a north-south axis; routes may be characterized by one or more bends or they may be looped.

II. Migration recording techniques

A. Field-glass observation

Various techniques of direct visual observation yield only partial estimates of bird migration.

B. Radar

Various radar tracking techniques provide detailed information on volume, direction, altitude, speed and composition of migration, and correlation with meteorological variables, in some instances in large-scale operations covering several thousand kilometres.

C. Radio tracking

Radio tracking provided insight into the migratory behaviour of single migrants for limited parts of their migratory flights.

III. Laboratory experiments

A. Basic sensory capabilities

1. Sensitivity to ambient pressure changes

Pigeons are sensitive to pressure changes of at least 10 mm H_2O, possibly 1 mm H_2O within 10 s. This capacity may be used as a "physiological barometer".

2. Hearing—infrasound

Pigeons at least hear infrasound possibly down to 0·06 Hz. This capacity might be used to locate sources of infrasound over very large distances.

3. Olfaction

In laboratory experiments response to laboratory chemicals only has been obtained, though the use of olfaction for foraging has been demonstrated in some species.

4. Magnetoreception

See Chapter III, Section B, 4.

5. Vision

The visual spectrum of pigeons, at least, seems to extend into the ultraviolet range; they are also able to discriminate the plane of polarized light. Visual acuity appears to be in the range of 1·94′ to 4·00′.

B. Integrated sensory capabilities

1. Image vision

Image vision, as used to recognize familiar landmarks, does not seem to be strongly developed in pigeons.

2. Sun compass

By allowing for the sun's apparent movement (in azimuth), birds use the sun for compass orientation. The sun compass of northern-hemisphere birds works satisfactorily with $3°-5°$ accuracy in the northern hemisphere, but no experiments in the southern hemisphere with transequatorial migrants have been carried out.

3. The star compass

Portions of the starry sky may be used by migratory birds for star compass orientation. Birds may be trained to respond to completely artificial star patterns.

4. The magnetic compass

The earth's magnetic field can be used for compass purposes. The birds do not use the polarity of the field, but seem to evaluate the inclination of the field lines to the vertical. Avian magnetoreception appears to be adaptable to different intensity ranges and sensitive to changes of 0·005 Oe or less. There are as yet only theoretical models of the physiological mechanism of magnetoreception.

5. Interrelation or hierarchy of compasses

The basic avian compass may be the magnetic compass; it is possible that the sun compass and the star compass are aligned or calibrated relative to the magnetic compass.

6 Compass and distance

Information on distance as well as direction may enable a bird to reach a goal so specified. The information on distance may be provided as time or amount of activity to be spent in migration. Even bent or looped migratory pathways may be explained by mechanisms of this type.

IV. Experimental field work with wild birds

A. Displacements from breeding or wintering ranges

Adult birds taken from their nests have been shown to home over distances of up to 6600 km. Return to winter ranges within the same winter, after displacement up to 260 km away, or for the next winter after displacement up to 3860 km away, has also been recorded.

B. Displacements during migration

Juvenile birds may perform their first autumn migration by direction-and-distance migration (vector navigation) as outlined in Chapter III, Section B, 6 and Chapter VI, Section G.

C. Establishment of breeding and of wintering ranges

Breeding and wintering ranges initially reached by juvenile birds by vector navigation, seem to be navigationally established for subsequent migration by a process that may be called "navigational imprinting", taking place within a short period before departure from the area of fledging and a short period after arrival in the winter area.

D. Initial orientation of displaced birds

When displaced and released, some birds such as mallards show some, possibly population-specific, stereotype initial orientation, the so-called "nonsense orientation".

E. Direct experimental interactions

"Nonsense orientation" is based on the sun compass during daytime and stars are probably used at night. Evidence of the use of olfactory cues in homing to nest sites has been reported in some but not all species tested.

V. Homing experiments with pigeons

Homing of the domestic homing pigeon is clearly superior to that of its ancestor, the wild rock pigeon.

A. Sun compass

The sun compass is used by homing pigeons during initial orientation. If the sun is not available, alternative mechanisms take over.

B. Magnetic cues

During initial orientation pigeons may also use their magnetic compass. They react to "magnetic storms". Young pigeons seem to calibrate their sun compass by a learning process on the basis of their magnetic compass. Pigeons possibly also use the geomagnetic field for navigational information.

C. Gravitation

Fluctuations of initial orientation seem to be correlated with lunar month, possibly related to corresponding fluctuations of gravity.

D. Visual cues

Visual cues, certainly image vision, do not seem to be essential for homing to the immediate vicinity of the loft. The final approach to the loft does require vision.

E. Olfactory navigation

There is some experimental support for the hypothesis that environmental olfactory cues may be used for navigation.

F. Map and compass

The concept of map and compass is still a frame for the phenomenon of migration though the properties of the map remain largely unspecified.

G. Vector navigation

The concept of vector navigation involving information on direction and distance is supported by experimental results. This system, which is of limited scope, may explain how juvenile birds accomplish their first autumn migration.

IX.   The use of statistics in this book

Few data collected in orientation research are so clear-cut as to require no statistical treatment. One such experimental example is the initial orientation of clock-shifted birds, as compared to that of control birds (e.g. Fig. 45). Most data are from two-dimensional distributions and many can be reduced to circular distributions (on a unit circle), but "linear" statistics cannot be used to analyse these.

Relevant statistical procedures for the analysis of circular or two-dimensional data have now been gathered from the literature, and new ones have been developed and are being developed for the special needs of orientation research (e.g. Watson and Williams, 1956; Schmidt-Koenig, 1961, 1965, 1975; Watson, 1961, 1962; Batschelet, 1965, 1972, 1978; Mardia, 1972).

Statistical tests have been applied routinely to the data reviewed in this book, either originally by the authors, or in the case of either no treatment or some inadequate treatment they have been carried out on data included in this book. The tests used were the Rayleigh test for uniformity (i.e. whether or not a sample is "random"); the V-test for directedness into a specified direction such as, e.g. the home direction; the Watson and Williams test for directional differences between two samples; the Watson-U^2 test for differences between two samples; the Mann-Whitney-U test for differences in homing performance; and regression and correlation statistics in appropriate cases.

Second-order statistics play an important role in orientation research. Orientational reactions of animals or experimental effects are frequently rather weak, i.e. they have a large scatter so that small differences between experimental and control birds, or some theoretical expectation may be masked, or require very large samples, or second-order statistics may be needed. Sometimes, raw data do not meet the requirements of independence and second-order statistics may be used to help overcome this difficulty.

Wiltschko's experiments with robins may be cited as an example. Several hundred to several thousand single directional recordings (made in one night), of a migratorily restless robin, are reduced by vector addition to one mean value. The mean values of several birds and nights then represent a second-order circular distribution processed statistically and interpreted by the experimenter. Second-order statistics have the potential to detect weak but consistent features, for example orientation to magnetic fields.

References

Able, K. P. (1972). Fall migration in coastal Louisiana and the evolution of migration patterns in the Gulf region. *Wilson Bull.* **84**, 231–242.

Able, K. P. (1973). The role of weather variables and flight direction in determining the magnitude of noctural bird migration. *Ecology* **54**, 1031–1041.

Able, K. P. (1974). Environmental influences on the orientation of free-flying nocturnal bird migrants. *Anim. Behav.* **22**, 224–238.

Able, K. P. (1978). Field studies of the orientation cue hierarchy of nocturnal songbird migrants. *In* "Animal Migration, Navigation and Homing" (K. Schmidt-Koenig and W. T. Keeton, Eds). Proceedings in Life Sciences, pp. 228–238. Springer Verlag, Berlin, Heidelberg, New York.

Adler, H. E. (1963). Psychological limits of celestial navigation hypotheses. *Ergeb. Biol.* **26**, 235–252.

Adler, H. E. (1971). Ed. "Orientation: Sensory Basis" Vol. 188. N.Y. Academy of Science, New York.

Adler, H. E. and Dalland, I. J., (1959). Spectral thresholds in the starling (*Sturnus vulgaris*). *J. Comp. Physiol. Psychol.* **52**, 438–445.

Alberts, J. R. and Friedman, M. I. (1972). Olfactory bulb removal but not anosmia increases emotionality and mouse killing. *Nature* **238**, 454–455.

Alerstam, T. (1976). Bird migration in relation to wind and topography. Dissertation, University of Lund.

Alexander, J. R. (1975). The effect of various phase-shifting experiments on homing in pigeons. Ph.D. Thesis, Cornell Univ., Ithaca, New York.

Alexander, J. R. and Keeton, W. T. (1972). The effect of directional training on initial orientation in pigeons. *Auk* **89**, 280–298.

Alexander, J. R. and Keeton, W. T. (1974). Clock-shifting effect on initial orientation of pigeons. *Auk* **91**, 370–374.

Alleva, E., Baldaccini, N. E., Foà, A. and Visalberghi, E. (1975). Homing behaviour of the rock pigeon. *Monitore Zool. Ital.* (N.S.) **9**, 213–224.

Arnould-Taylor, W. E. and Malewski, A. N. (1955). The factor of topography in bird homing experiments. *Ecology* **36**, 641–646.

Baldaccini, N. E., Benvenuti, S., Fiaschi, V. and Papi, F. (1975). Pigeon navigation: effects of wind deflection at home cage on homing behaviour. *J. Comp. Physiol.* **99**, 177–186.

Bang, B. G. (1971). Functional anatomy of the olfactory system in 23 orders of birds. *Acta Anat. Suppl.* **58**, 1–76.

Bang, B. G. and Cobb, S. (1968). The size of the olfactory bulb in 108 species of birds. *Auk* **85**, 55–61.

Barlow, J. S. (1964). Inertial navigation as a basis for animal navigation. *J. Theoret. Biol.* **6**, 76–117.

Barlow, J. S. (1966). Inertial navigation in relation to animal navigation. *J. Inst. Navigation* **19**, 302–316.

Batschelet, E. (1965). "Statistical methods for the Analysis of Problems in Animal Orientation and Certain Biological Rhythms." The American Institute of Biological Sciences, Washington, D.C.

Batschelet, E. (1972). Recent statistical methods for orientation data. *In* "Animal Orientation and Navigation." (S. R. Galler, K. Schmidt-Koenig, G. J. Jacobs and R. E. Belleville, Eds) pp. 61–91. NASA SP–262 US Govt. Printing Office, Washington D.C.

Batschelet, E. (1978). Second-order statistical analysis of directions. *In* "Animal Migration, Navigation and Homing". (K. Schmidt-Koenig and W. T. Keeton, Eds) pp. 3–24. Proceedings in Life Sciences. Springer Verlag, Berlin, Heidelberg, New York.

Beaugrand, J. P. (1976). An attempt to confirm magnetic sensitivity in the pigeon, *Columba livia*. *J. Comp. Physiol.* **110**, 343–355.

Becker, G. (1963). Magnetfeld-Orientierung von Dipteren. *Naturwiss.* **50**, 664.

Bellrose, F. C. (1958). The orientation of displaced waterfowl in migration. *Wilson Bull.* **70**, 20–40.

Bellrose, F. C. (1971). The distribution of nocturnal migrants in the air space. *Auk* **88**, 397–424.

Bellrose, F. C. and Graber, R. R. (1963). A radar study of the flight directions of nocturnal migrants. *Proc. XIII. Int. Ornithol. Congr.* Ithaca N.Y., 362–389.

Benvenuti, S., Fiaschi, V., Fiore, L. and Papi, F. (1973a). Homing performances of inexperienced and directionally trained pigeons subjected to olfactory nerve section. *J. Comp. Physiol.* **83**, 81–92.

Benvenuti, S., Fiaschi, V., Fiore, L. and Papi, F. (1973b). Disturbances of homing behaviour in pigeons experimentally induced by olfactory stimuli. *Monitore Zool. Ital.* (N.S.) **7**, 117–128.

Benvenuti, S., Fiaschi, V. and Foà, A. (1977). Homing behaviour of pigeons disturbed by application of an olfactory stimulus. *J. Comp. Physiol.* **120**, 173–179.

Berthold, P. (1973). Relationship between migratory restlessness and migration distances in six *Sylvia* species. *Ibis* **155**, 594–599.

Berthold, P. (1978). Concept of endogenous control of migration in warblers. *In* "Animal Migration, Navigation and Homing". (K. Schmidt-Koenig and W. T. Keeton, Eds) pp. 275–282. Proceedings in Life Sciences. Springer Verlag, Berlin, Heidelberg, New York.

Berndt, R. and Winkel, W. (1978). Field experiments on problems of imprinting to the birthplace in the Pied Flycatcher *Ficedula hypoleuca*. XVII International Ornithological Congress, Berlin.

Berthold, P., Gwinner, E., Klein, H. and Westrich, P. (1972). Beziehungen zwischen Zugunrube und Zugablauf bei Gartenund Mönchsgrasmücke (*Sylvia borin* und *S. atricapilla*). *Z. Tierpsychol.* **30**, 26–35.

Billings, S. M. (1968). Homing in Leach's petrel. *Auk* **85**, 36–43.

Birner, M., Gernandt, D., Merkel, F. W. and Wiltschko, W. (1968). Verfrachtungsversuche mit einer Starenpopulation im Winter. *Natur Mus.* **98**, 507–514.

Blough, D. S. (1956). Dark adaptation in the pigeon. *J. Comp. Physiol. Psychol.* **49**, 425–430.

Blough, D. S. (1957). Spectral sensitivity in the pigeon. *J. Opt. Soc. Amer.* **47**, 827–833.

Blough, P. M. (1971). The visual acuity of the pigeon for distant targets. *J. Exp. Anal. Behav.* **15**, 57–67.

Bookman. M. A. (1978). Sensitivity of the homing pigeon to an earth-strength magnetic field. *In* "Animal Migration, Navigation and Homing". (K. Schmidt-Koenig and W. T. Keeton, Eds) pp. 127–134. Proceedings in Life Sciences. Springer Verlag, Berlin, Heidelberg, New York.

Bruderer, B. (1977). Beitrag der Radar-Ornithologie zu Fragen der Orientierung, der Zugphysiologie und der Umweltabhängigkeit des Vogelzuges. *Vogelwarte* (Sonderheft) **29**, 83–91.

Bruderer, B. (1978). Effects of alpine topography and winds on migrating birds. In "Animal Migration, Navigation and Homing". (K. Schmidt-Koenig and W. T. Keeton, Eds) pp. 252–265. Proceedings in Life Sciences. Springer Verlag, Berlin, Heidelberg, New York.

Bruderer, B. and Steidinger, P. (1972). Methods of quantitative and qualitative analysis of bird migration with a tracking radar. *In* "Animal Orientation and Navigation". (S. R. Galler, K. Schmidt-Koenig, G. J. Jacobs and R. E. Belleville, Eds) pp. 151–167. NASA SP–262 US Govt. Printing Office, Washington D.C.

Bruderer, B. and Weitnauer, E. (1972). Radarbeobachtungen über Zug und Nachtflüge des Mauerseglers (*Apus apus*). *Rev. Suisse Zool.* **79**, 1190–1200.

Bruderer, B. and Winkler, R. (1976). Vogelzug in den schweitzer Alpen. Eine Ubersicht über Entwicklung und Stand der Forschung. *Angew. Ornithol.* **5**, 32–55.

Catania, A. C. (1961). Techniques for the control of monocular and binocular viewing in the pigeon. *J. Exp. Anal. Behav.* **6**, 627–629.

Catania, A. C. (1964). On the visual acuity of the pigeon. *J. Exp. Anal. Behav.* **7**, 361–366.

Chapin, J. P. (1932). The birds of the Belgian Congo. *Bull. Amer. Mus. Nat. Hist.* **65**, Part I. 322–362.

Chard, R. D. (1939). Visual acuity in the pigeon. *J. Exp. Psychol.* **24**, 588–608.

Chelazzi, G. and Pineschi, F. (1974). Recoveries after displacement in pigeons belonging to an urban population. *Monitore Zool. Ital.* (N.S.) **8**, 151–157.

Clark, C. L., Peck, R. A. and Hollander, W. F. (1948). Homing pigeon in electromagnetic fields. *J. Appl. Phys.* **19**, 1183.

Cochran, W. W. (1972). Long-distance tracking of birds. *In* "Animal Orientation and Navigation". (S. R. Galler, K. Schmidt-Koenig, G. J. Jacobs and R. E. Belleville, Eds) pp. 39–59. NASA SP–262 US Govt. Printing Office, Washington D.C.

Cochran, W. W., Montgomery, G. G. and Graber, R. R. (1967). Migratory flights of Hylocichla-thrushes in spring: a radio telemetry study. *The Living Bird* **6**, 213–225.

Davis, L. (1948). Remarks on: "The physical basis of bird navigation". *J. Appl. Phys.* **19**, 307–308.

Decoursey, P. J. (1962). Effect of light on the circadian activity rhythm of the flying squirrel (*Glaucomys volans*). *Z. Vergl. Physiol.* **44**, 331–354.

Delius, J. D. and Emmerton, J. (1978). Sensory mechanisms related to homing in pigeons. *In* "Animal Migration, Navigation and Homing". (K. Schmidt-Koenig and W. T. Keeton, Eds) pp. 35–41. Proceedings in Life Sciences. Springer Verlag, Berlin, Heidelberg, New York.

Delius, J. D., Perchard, R. J. and Emmerton, J. (1976). Polarized light discrimination by pigeons and an electroretinographic correlate. *J. Comp. Physiol. Psychol.* **90**. 560–571.

Dementiew, G. P. and Gladkow, N. A. (1954). "Die Vögel der Sowjetunion." Moskau.

Dobben, W. H. van. (1939). Stichting Vogeltrekstation Texel. Jaarverslag, 1937 en 1938, pp. 12–14.

Donner, K. O. (1951). The visual acuity of some passerine birds. *Acta Zool. Fennica* **66**. 1–40.

Dorst, J. (1962). "The Migration of Birds." Heinemann Ltd., London.

Downhower, I. F. and Windsor, D. M. (1971). Use of landmarks in orientation by bank swallows. *BioScience* **21**. 570.

Drury, W. H. and Keith, J. A. (1962). Radar studies of songbird migration in coastal New England. *Ibis* **104**, 449–489.

Drury, W. H. and Nisbet, I. C. T. (1964). Radar studies of orientation of songbird migrants in south-eastern New England. *Bird-Banding* **35**, 69–119.

Eastwood, E. (1967). "Radar Ornithology." Methuen, London.

Edrich, W. and Keeton, W. T. (1977). A comparison of homing behaviour in feral and homing pigeons. *Z. Tierpsychol.* **44**. 389–401.

Edrich, W. and Keeton, W. T. (1978). Further investigations of the effect of "Flight during clock-shift" on pigeon orientation. *In* "Animal Migration, Navigation and Homing". (K. Schmidt-Koenig and W. T. Keeton, Eds) pp. 184–193. Proceedings in Life Sciences. Springer Verlag, Berlin, Heidelberg. New York.

Elsner, B. (1978). Accurate measurements of the initial track (radius 1500 m) of homing pigeons. *In* "Animal Migration, Navigation and Homing". (K. Schmidt-Koenig and W. T. Keeton, Eds) pp. 194–198.

Proceedings in Life Sciences. Springer Verlag, Berlin, Heidelberg, New York.

Emlen, S. T. (1967a). Migratory orientation in the Indigo Bunting, *Passerina cyanea*. Part I: The evidence for use of celestial cues. *Auk* **84**, 309–342.

Emlen, S. T. (1967b). Migratory orientation in the Indigo Bunting, *Passerina cyanea*. Part II: Mechanisms of celestial orientation. *Auk* **84**, 463–489.

Emlen, S. T. (1967c). Orientation of Zugunruhe in the Rose-breasted Grosbeak, *Pheucticus ludovicianus*. *Condor* **69**, 203–205.

Emlen, S. T. (1970). The influence of magnetic information on the orientation of the indigo bunting, *Passerina cyanea*. *Anim Behav*. **18**, 215–225.

Emlen, S. T. (1971). Unconventional theories of orientation; panel discussion. *Ann. N.Y. Acad. Sci*. **188**, 331–359.

Emlen, S. T. (1972). The ontogenetic development of orientation capabilities. *In* "Animal Orientation and Navigation". (S. R. Galler, K. Schmidt-Koenig, G. J. Jacobs and R. E. Belleville, Eds) pp. 191–210. NASA SP-262 US Govt. Printing Office, Washington D.C.

Emlen, S. T. (1974). Problems of identifying bird species by radar signature analyses; intra-specific variability. *In* "The biological aspects of the bird/aircraft collision problem". (S. Gauthreaux, Ed.) pp. 509–524. Air Force Office of Scientific Research, Clemson N.C., USA.

Emlen, S. T. (1975). Migration: orientation and navigation. *In* "Avian Biology". (D. S. Farner and J. R. King, Eds) vol. V. pp. 129–219. Academic Press, New York and London.

Emlen, S. T. and Demong, N. J. (1978). Orientation strategies used by free-flying bird migrants: a radar tracking study. *In* "Animal Migration, Navigation and Homing". (K. Schmidt-Koenig and W. T. Keeton, Eds) pp. 283–293. Proceedings in Life Sciences. Springer Verlag, Berlin, Heidelberg, New York.

Emlen, S. T. and Emlen, J. T. (1966). A technique for recording migratory orientation of captive birds. *Auk* **83**, 361–367.

Enright, J. T. (1972). A virtuose isopod: circa-lunar rhythms and their tidal fine structure. *J. Comp. Physiol*. **77**, 141–162.

Exner, S. (1893. Negative Versuchsergebnisse über das Orientierungsvermögen von Brieftauben. *Sber. Akad. Wiss. Wien*, Abt. 1. **102**, 318–331.

Fell, H. B. (1947). Migration of *Chalcites lucidus* of New Zealand. *Transact. Proc. Royal Soc. N.Z*. **76**, 504–515.

Fiaschi, V. and Wagner, G. (1976). Pigeon homing: some experiments for testing the olfactory hypothesis. *Experientia* **32**, 991–993.

Fiaschi, V., Farini, A. and Ioalé, P. (1974). Homing experiments on swifts *Apus apus* (L.) deprived of olfactory perception. *Monitore Zool. Ital.* N.S.) **8**, 235–244.

Frisch, K. von (1968). "Dance, Language and Orientation of Bees." Harvard Univ. Press, USA.

Fromme, H. G. (1961). Untersuchungen über das Orientierungsvermögen nächtlich ziehender Kleinvögel, Erithacus rubecula, *Sylvia communis*. *Z. Tierpsychol*. **18**, 205–220.

Galifret, Y. (1968). Les diverses aires fonctionelles de la rétine du pigeon. *Z. Zellforsch.* **86**, 535–545.

Galler, S. R., Schmidt-Koenig, K., Jacobs, G. J. and Belleville, R. E. (1972). "Animal Orientation and Navigation." NASA SP-262. US Govt. Printing Office, Washington D.C.

Gauthreaux, S. A. (1972). A radar and direct visual study of passerine spring migration in southern Louisiana. *Auk* **88**, 343–365.

Gauthreaux, S. A. (1978). Importance of the daytime flights of nocturnal migrants: redetermined migration following displacement. *In* "Animal Migration, Navigation and Homing". (K. Schmidt-Koenig and W. T. Keeton, Eds) pp. 219–227. Proceedings in Life Sciences. Springer Verlag, Berlin, Heidelberg, New York.

Gehring, W. (1963). Radar- und Feldbeobachtungen über den Verlauf des Vogelzuges im Schweizerischen Mittelland: Der Tageszug im Herbst (1957–1961). *Ornithol. Beob.* **60**, 35–68.

Goodloe, L. (1974). Night homing in pigeons. Ph. D. Thesis, Cornell Univ., Ithaca, N.Y., USA.

Gordon, D. A. (1948). Sensitivity of the homing pigeon to the magnetic field of the earth. *Science* **108**, 710–711.

Graber, R. R. (1965). Night flight with a thrush. *Audubon Mag.* **67**, 368–374.

Graue, L. C. (1963). The effect of phase shifts in the day-night cycle on pigeon homing at distances of less than one mile. *Ohio J. Sci.* **63**, 214–217.

Graue, L. C. (1965). Experience effect on initial orientation in pigeon homing. *Anim. Behav.* **13**, 149–153.

Graue, L. C. (1970). Orientation and distance in pigeon homing (*Columba livia*). *Anim. Behav.* **18**, 36–40.

Graue, L. C. and Pratt, J. G. (1959). Directional differences in pigeon homing in Sacramento, California and Cedar Rapids, Iowa. *Anim. Behav.* **7**, 201–208.

Griffin, D. R. (1940). Homing experiments with Leach's petrels. *Auk* **57**, 61–74.

Griffin, D. R. (1943). Homing experiments with herring gulls and common terns. *Bird-Banding* **14**, 7–33.

Griffin, D. R. (1952). Bird navigation. *Biol. Rev. Cambridge Phil. Soc.* **27**. 359–400.

Griffin, D. R. (1955). Bird navigation. *In* "Recent Studies in Avian Biology". (A. Wolfson, Ed.) pp. 154–197. Univ. of Illinois Press, Urbana, USA.

Griffin, D. R. (1969). The physiology and geophysics of bird navigation. *Quart. Rev. Biol.* **4**, 255–276.

Griffin, D. R. (1972). Nocturnal bird migration in opaque clouds. *In* "Animal Orientation and Navigation". (S. R. Galler, K. Schmidt-Koenig, G. J. Jacobs and R. E. Belleville, Eds) pp. 169–188. NASA SP-262 Govt. Printing Office, Washington D.C., USA.

Griffin, D. R. (1973). Oriented bird migration in or between opaque cloud layers. *Proc. Amer. Phil. Soc.* **117**, 117–141.

Gronau, J. (1971). Spezielle Untersuchungen zur jahreszeitlichen Variabilität des Orientierungsverhaltens unerfahrener und erfahrener Reisebrieftauben unter besonderer Berücksichtigung neuerer methodisch-statistischer Aspekte. Dissertation, Universität Göttingen.

Gronau, J. and Schmidt-Koenig, K. (1970). Annual fluctuation in pigeon homing. *Nature, Lond.* **226**, 87–88.

Grubb, T. C. (1971). Olfactory navigation by Leach's petrel and other procellariiform birds. Ph. D. Thesis, University of Wisconsin, Madison, USA.

Grubb, T. C. (1972). Smell and foraging in shearwaters and petrels. *Nature, Lond.* **237**, 404–405.

Gualtierotti, T., Schreiber, B., Mainardi, D. and Passerini, D. (1959). Effect of acceleration on cerebellar potentials in birds and its relation to sense of direction. *Amer. J. Physiol.* **197**, 469–474.

Gwinner, E. (1968a). Artspezifische Muster der Zugunruhe bei Laubsängern und ihre mögliche Bedeutung für die Beendigung des Zuges im Winterquartier. *Z. Tierpsychol.* **25**, 843–853.

Gwinner, E. (1968b). Circannuale Periodik als Grundlage des jahreszeitlichen Funktionswandels bei Zugvögeln. Untersuchungen am Fitis (*Phylloscopus trochilus*) und am Waldlaubsänger (*P. sibilatrix*). *J. Ornithol.* **109**, 70–95.

Gwinner, E. (1969). Untersuchungen zur Jahresperiodik von Laubsängern. *J. Ornithol.* **110**, 1–21.

Gwinner, E. (1972). Endogenous timing factors in bird migration. *In* "Animal Orientation and Navigation". (S. R. Galler, K. Schmidt-Koenig, G. J. Jacobs and P. E. Belleville, Eds) pp. 321–338. NASA SP-262 US Govt. Printing Office, Washington D.C., USA.

Gwinner, E. (1974). Endogenous temporal control of migratory restlessness in warblers. *Naturwiss* **61**, 405–406.

Gwinner, E. and Wiltschko, W. (1978). Endogenously controlled changes in migratory direction of the garden warbler, *Sylvia borin*. *J. Comp. Physiol.* **125**, 267–273.

Haase, E., Otto, C. and Murbach, H. (1977). Brain weight in homing and "non-homing" pigeons. *Experientia* **33**, 606.

Hachet-Souplet, P. (1911). L'instinct du retour chez le pigeon voyageur. *Rev. Sci.* **29**, 231–238.

Hamilton, W. J. (1962a). Celestial orientation in juvenal waterfowl. *Condor* **64**, 19–33.

Hamilton, W. J. (1962b). Bobolink migratory pathways and their experimental analysis under night skies. *Auk* **79**, 208–233.

Hamilton, W. J. (1966)). Anaylsis of bird navigation experiments. *In* "Systems Analysis in Ecology". (K. E. F. Watt, Ed.) pp. 174–178. Academic Press, New York and London.

Hamilton, W. and Goldstein, J. L. (1933). Visual acuity and accomodation in the pigeon. *J. Comp. Psychol.* **15**, 193–197.

Hartwick, R. F., Foà, A. and Papi, F. (1977). The effect of olfactory deprivation by nasal tubes upon homing behaviour in pigeons. *Behav. Ecol. Sociobiol.* **2**, 81–84.

Hartwick, R. F., Kiepenheuer, J. and Schmidt-Koenig, K. (1978). Further experiments on the olfactory hypothesis of pigeon navigation. *In* "Animal Migration, Navigation and Homing". (K. Schmidt-Koenig and W. T. Keeton, Eds) pp. 107–118. Proceedings in Life Sciences, Springer Verlag, Berlin, Heidelberg, New York.

Hasler, A. D. and Scholz, A. T. (1978). Olfactory imprinting in Coho Salmon (*Oncorhynchus kisutch*). *In* "Animal Migration, Navigation and Homing". (K. Schmidt-Koenig and W. T. Keeton, Eds) pp. 356–369. Proceedings in Life Sciences. Springer Verlag, Berlin, Heidelberg, New York.

Hasler, A. D. and Wisby, W. J. (1951). Discrimination of stream odours by fishes and relation to parent stream behaviour. *Amer. Nat.* **85**, 223–238.

Henton, W. W. (1966). Suppression behaviour to odorous stimuli in the pigeon. Ph. D. Thesis, Florida State Univ., Tallahassee.

Henton, W. W. (1969). Conditioned suppression to odorous stimuli in pigeons. *J. Exp. Anal. Behav.* **12**. 175–185.

Henton, W. W., Smith, J. C. and Tucker, D. (1966) Odour discrimination in pigeons. *Science* **153**, 1138–1139.

Hitchcock, H. B. (1952). Aeroplane observations of homing pigeons. *Proc. Amer. Phil. Soc.* **96**, 270–289.

Hitchcock, H. B. (1955). Homing flights and orientation in pigeons. *Auk* **72**, 355–373.

Hoffmann, K. (1953). Die Einrechnung der Sonnenwanderung bei der Richtungsweisung des sonnenlos aufgezogenen Stares. *Naturwiss,* **40**, 148.

Hoffmann, K. (1954). Versuche zu der im Richtungsfinden der Vögel enthaltenen Zeitschätzung. *Z. Tierpsychol* **11**, 453–475.

Hoffmann, K. (1958). Repetition of an experiment on bird orientation. *Nature* **181**, 1435–1437.

Hoffmann, K. (1959). Die Richtungsorientierung von Staren unter der Mitternachtssonne. *Z. Vergl. Physiol,* **41**, 471–480.

Huizinger, E. (1935). Durchschneidung aller Bogengänge dei der Taube. *Pfluegers Arch. Ges. Physiol.* **236**, 52–58.

Idler, D. R., McBride, J. R., Jonas, R. E. and Tomlinson, N. (1961). Olfactory perception in migrating salmon. II. Studies on a laboratory bioassey for home-stream water and mammalian repellant. *Can. J. Biochem. Physiol.* **39**, 1575–1584.

Ising, G. (1945). Die physikalische Möglichkeit eines tierischen Orientierungssinnes auf der Basis der Erdrotation. *Ark. Mat. Astron. Fys.* **32**, 1–23.

Keeton, W. T. (1969). Orientation by pigeons: is the sun necessary? *Science* **165**, 922–928.

Keeton, W. T. (1970a). "Distance effect" in pigeon orientation: An evaluation. *Biol. Bull,* 139–519.

Keeton, W. T. (1970b). Orientation by pigeons. *Science* **168**, 153.

Keeton, W. T. (1970c). Do pigeons determine latitudinal displacement from the sun's altitude? *Nature, Lond.* **227**, 626–627.

Keeton, W. T. (1971a). Magnets interfere with pigeon homing. *Proc. Nat. Acad. Sci. USA* **68**, 102–106.

Keeton, W. T. (1971b). Remarks in a panel discussion: Unconventional theories of orientation. *Ann. N.Y. Acad. Sci.* **188**, 331–333, 338–340.

Keeton, W. T. (1972). Effects of magnets on pigeon homing. *In* "Animal Orientation and Navigation". (S. R. Galler, K. Schmidt-Koenig, G. J. Jacobs and R. E. Belleville, Eds) pp. 579–594. NASA SP-262 US Govt. Printing Office, Washington D.C.

Keeton, W. T. (1973). Release-site bias as a possible guide to the "map" component in pigeon homing. *J. Comp. Physiol.* **86**, 1–16.

Keeton, W. T. (1974a). The navigational and orientational basis of homing in birds. *In* "Advances in Study behavior". Vol. 5, pp. 47–132. Academic Press, New York and London.

Keeton, W. T. (1974b). The mystery of pigeon homing. *Scient. Amer.* **23**, 96–107.

Keeton, W. T. (1974c). Pigeon homing: no influence of outward-journey detours on initial orientation. *Monitore Zool. Ital.* (N.S.) **8**, 227–234.

Keeton, W. T. and Alexander, J. R. (1978). The effect of exercise flights during phaseshifting on the orientation of homing pigeons. (In press.)

Keeton, W. T. and Brown, A. I. (1976). Homing behaviour of pigeons not disturbed by application of an olfactory stimulus. *J. Comp. Physiol*, **105**, 259–266.

Keeton, W. T., Larkin, T. S. and Windsor, D. M. (1974). Normal fluctuations in the earth's magnetic field influence pigeon orientation. *J. Comp. Physiol.* **95**, 95–103.

Keeton, W. T., Kreithen M. L. and Hermayer, K. L. (1977). Orientation by pigeons deprived of olfaction by nasal tubes. *J. Comp. Physiol.* **114**, 289–299.

Kenyon, K. W. and Rice, D. W. (1958). Homing of Laysan Albatrosses. *Condor* **60**, 3–6.

Kiepenheuer, J. (1978a). Pigeon homing: A repetition of the deflector loft experiment. *Behav. Ecol Sociobiol.* **3**, 393–395.

Kiepenheuer, J. (1978b). Pigeon navigation and magnetic field: Information collected during the outward journey is used in the homing process. *Naturwiss.* **65**, 113.

Kiepenheuer, J. (1978c). Inversion of the Magnetic field during transport: its influence on the homing behaviour of pigeons. *In* "Animal Migration, Navigation and Homing". (K. Schmidt-Koenig and W. T. Keeton, Eds) pp. 135–142. Proceedings in Life Sciences. Springer Verlag, Berlin, Heidelberg, New York.

Kiepenheuer, J. (1978d). The importance of outward journey information in pigeon homing. *Int. XVII. Ornithol. Congr.*, Berlin 1978.

Kiepenheuer, J. (1979). Pigeon homing: Deprivation of olfactory information does not affect the deflector effect. (Submitted to *Behavioural Processes*.)

Klomp, H. (1949). Vogeltrekstation Texel. Jaarverslag 1948, 2–5.

Klomp, H. (1950). Orientatieproeven. Vogeltrekstation Texel. Jaarverslag 1949, 3–5.

Kluijver, H. W. (1935). Ergebnisse eines Versuches über das Heimfindevermögen von Staren. *Ardea* 24, 227–239.

Köhler, K. L. (1978). Do pigeons use their eyes for navigation? A new technique! *In* "Animal Migration, Navigation and Homing". (K. Schmidt-Koenig and W. T. Keeton, Eds) pp. 57–64. Proceedings in Life Sciences. Springer Verlag, Berlin, Heidelberg, New York.

Kramer, G (1931). Zug in großer Höhe. *Vogelzug* 2, 69–71.

Kramer, G. (1948). Neue Beiträge zur Frage der Fernorientierung der Vögel. *Orn. Ber.* 1, 228–239.

Kramer, G. (1949). "Über Richtungstendenzen bei der nächtlichen Zugunruhe gekäfigter Vögel. Ornithologie als biologische Wissenschaft." (E. Mayr and E. Schüz, Eds) Heidelberg.

Kramer, G. (1950a). Orientierte Zugaktivität gekäfigter Singvögel. *Naturwiss.* 37, 188.

Kramer, G. (1950b). Weitere Analyse der Faktoren, welche die Zugaktivität des gekäfigten Vogels orientieren. *Naturwiss* 37, 377–378.

Kramer, G. (1951). Eine neue Methode zur Erforschung der Zugorientierung und die bisher damit erzielten Ergebnisse. *Proc. X. intern. Ornithol. Congr.* Uppsala 1950, 269–280.

Kramer, G. (1953). Wird die Sonnenhöhe bei der Heimfindeorientierung verwertet? *J. Ornithol.* 94, 201–219.

Kramer, G. (1955). Ein weiterer Versuch, die Orientierung von Brieftauben durch jahreszeitliche Änderung der Sonnenhöhe zu beeinflussen. Gleichzeitig eine Kritik der Theorie des Versuches. *J. Ornithol.* 96, 173–185.

Kramer, G. (1957). Experiments on bird orientation and their interpretation. *Ibis* 99. 196–227.

Kramer, G. (1959). Recent experiments on bird orientation. *Ibis* 101, 399–416.

Kramer, G. and Riese, E. (1952). Die Dressur von Brieftauben auf Kompassrichtung im Wahlkäfig. *Z. Tierpsychol.* 9, 245–251.

Kramer, G. and St. Paul, U. von (1950a). Stare (*Sturnus vulgaris* L.) lassen sich auf Himmelsrichtung dressieren. *Naturwiss.* 37, 526–527.

Kramer, G. and St. Paul, U. von (1950b). Ein wesentlicher Bestandteil der Orientierung der Reisetaube: die Richtungsdressur. *Z. Tierpsychol.* 7, 620–631.

Kramer G. and St. Paul, U. von (1954). Das Heimkehrvermögen gekäfigter Brieftauben. *Ornithol. Beob.* 51, 3–12.

Kramer, G. and St. Paul, U. von (1956). Weitere Erfahrungen über den "Wintereffekt" beim Heimfindevermögen von Brieftauben. *J. Ornithol.* 97, 353–370.

Kramer, G., St. Paul, U. von and Wallraff, H. G. (1959). Über die Heimfindeleistung von unter Sichtbergrenzung aufgewachsenen Brieftauben. *Verh. Dt. Zool. Ges.* Frankfurt **1958**, 169–176.

Kreithen, M. L. (1978). Sensory mechanisms for animal orientation—can any new ones be discovered? *In* "Animal Migration, Navigation and Homing". K. Schmidt-Koenig and W. T. Keeton, Eds) pp. 25–34. Proceedings in Life Sciences. Springer Verlag, Berlin, Heidelberg, New York.

Kreithen, M. L. and Keeton, W. T. (1974a). Detection of changes in atmospheric pressure by the homing pigeon, *Columba livia*. *J. Comp. Physiol.* **89**, 73–82.

Kreithen, M. L. and Keeton, W. T. (1974b). Detection of polarized light by the homing pigeon, *Columba livia*. *J. Comp. Physiol.* **89**, 83–92.

Kreithen, M. L. and Keeton, W. T. (1974c). Attempts to condition homing pigeons to magnetic stimuli. *J. Comp. Physiol.* **91**, 355–362.

Kullenberg, B. (1946). Über Verbreitung und Wanderungen von vier Sterna-Arten. *Ark. Zool.* **38**.

Lack, D. (1960). The influence of weather on passerine migration: A review. *Auk* **77**, 171–209.

Lack, D. and Eastwood, E. (1962). Radar films of migration over eastern England. *British Birds* **55**, 388–414.

Larkin, T. S. and Keeton, W. T. (1976). Bar magnets mask the effect of normal magnetic disturbances on pigeon orientation. *J. Comp. Physiol.* **110**, 227–231.

Larkin, T. S. and Keeton, W. T. (1978). An apparent lunar rhythm in the day-to-day variations in initial bearings in homing pigeons. *In* "Animal Migration, Navigation and Homing". (K. Schmidt-Koenig and W. T. Keeton, Eds) pp. 92–106. Proceedings in Life Sciences. Springer Verlag, Berlin, Heidelberg, New York.

Leask, M. J. M. (1977). A physico-chemical mechanism for magnetic field detection by migratory birds and homing pigeons. *Nature, Lond.* **267**, 144–146.

Leask, M. J. M. (1978). Primitive models of magnetoreception. *In* "Animal Migration, Navigation and Homing". (K. Schmidt-Koenig and W. T. Keeton, Eds) pp. 318–322. Proceedings in Life Sciences. Springer Verlag, Berlin, Heidelberg, New York.

Levi, W. M. (1963). "The Pigeon." Levi Publ. Co., Sumter, S.C.

Lindauer, M. (1976). Orientierung der Tiere. Orientation in animals—old and new problems. *Verh. Dt. Zool. Ges.* **1976**, 156–183.

Lindauer, M. and Martin, H. (1968). Die Schwereorientierung der Bienen unter dem Einfluss des Erdmagnetfelds. *Z. Vergl. Physiol.* **60**, 219–243.

Lindauer, M. and Martin, H. (1972). Magnetic effect on dancing bees. *In* "Animal Orientation and Navigation". (S. R. Galler, K. Schmidt-Koenig, G. J. Jacobs and R. E. Belleville, Eds) pp. 559–567. NASA SP-262 US Govt. Printing Office, Washington D.C.

Lindley, S. B. (1930). The maze learning ability of anosmic and blind anosmic rats. *J. Genet. Psychol. Developm. Comp. Clin. Psychol.* **37**, 254-265.

Löhrl, H. (1959). Zur Frage des Zeitpunkts einer Prägung auf die Heimatregion beim Halsbandschnäpper (*Ficedula albicollis*). *J. Ornithol.* **100**, 132–140.

Lowery, G. H. (1951). A quantitative study of the nocturnal migration of birds. *Univ. Kans. Mus. Nat. Hist.* **3**, 361–472.

Lowery, G. H. and Newman, R. J. (1966). A continent-wide view of bird migration on four nights in October. *Auk* **83**, 547–586.

Mardia, K. V. (1972). Statistics of Directional Data." Academic Press, New York and London.

Marks, H. E., Remley, N. R., Seago, J. D. and Hastings, D. W. (1971). Effects of bilateral lesions of the olfactory bulbs of rats on measures of learning and motivation. *Physiol. Behav.* **7**, 1–6.

Marshall, A. J. (1956). The breeding cycle of the short-tailed shearwater in relation to trans-equatorial migration and its environment. *Proc. Zool. Soc.* London **127**, 489–510.

Martin, H. and Lindauer, M. (1977). Der Einfluß des Erdmagnetfeldes auf die Schwereorientierung der Honigbiene (*Apis mellifica*). *J. Comp. Physiol.* **122**, 145–187.

Masher, J. W. and Stolt, B. (1961). Lufttryckets inverkan pa ortolansparvens (*Emberiza hortulana* L.) aktivitet under varflyttningsperioded. *Var Fogelvarld* **20**, 97–111.

Matthews, G. V. T. (1951). The experimental investigation of navigation in homing pigeons. *J. Exp. Biol.* **28**, 508–536.

Matthews, G. V. T. (1953a). Navigation in the Manx shearwater. *J. Exp. Biol.* **30**, 370–396.

Matthews, G. V. T. (1953b). Sun navigation in homing pigeons. *J. Exp. Biol.* **30**, 243–267.

Matthews, G. V. T. (1953c). The orientation of untrained pigeons: a dichotomy in the homing process. *J. Exp. Biol.* **30**, 268–276.

Matthews, G. V. T. (1955). "Bird Navigation." Cambridge Univ. Press, Cambridge.

Matthews, G. V. T. (1961). "Nonsense" orientation in mallards (*Anas platyrhynchos*) and its relation to experiments on bird navigation. *Ibis* **103a**, 211–230.

Matthews, G. V. T. (1963a). "Nonsense" orientation as a population variant. *Ibis* **105**, 185–197.

Matthews, G. V. T. (1963b). The astronomical bases of "nonsense" orientation. *Proc. XIII. Int. Ornithol. Congr.* Ithaca, 415–429.

Matthews, G. V. T. (1963c). The orientation of pigeons as affected by the learning of landmarks and by the distance of displacement. *Anim. Behav.* **11**, 310–317.

Matthews, G. V. T. (1967). Some parameters of "nonsense" orientation in mallard. *Wildfowl Trust* **18**, 88–97.

Matthews, G. V. T. (1968). "Bird Navigation." 2nd. ed. Cambridge Univ. Press, Cambridge.

Matthews, G. V. T. (1971). "Vogelflug." W. Goldmann, München.

Matthews, G. V. T. (1974). On bird navigation, with some statistical undertones. *Ann. Royal Stat. Soc.* **197**, 349–364.

Mauersberger, G. (1957). Umsiedlungsversuche am Trauerschnäpper (*Muscicapa hypoleuca*), durchgeführt in der Sowjetunion. Ein Sammelreferat. *J. Ornithol.* **98**, 445–447.

McDonald, D. L. (1972). Some aspects of the use of visual cues in directional training of homing pigeons. *In* "Animal Orientation and Navigation." (S. R. Galler, K. Schmidt-Koenig, G. J. Jacobs and R. E. Belleville, Eds) pp. 293–304. NASA SP-262 US Govt. Printing Office, Washington D.C.

McDonald, D. L. (1973). The role of shadows in directional training and homing of pigeons, *Columba livia. J. Exp. Zool.* **183**, 267–280.

McClure, H. E. (1974). Migration and Survival of the Birds of Asia. US Army Medical Comp. SEATO Medical Project, Bangkok, Thailand.

Merkel, F. W. and Fromme, H. G. (1958). Untersuchungen über das Orientierungsvermögen nächtlich ziehender Rotkelchen, *Erithacus rubecula. Naturwiss.* **45**, 499–500.

Merkel, F. W. and Wiltschko, W. (1965). Magnetismus und Richtungsfinden zugunruhiger Rotkelchen (*Erithacus rubecula*). *Vogelwarte* **23**, 71–77.

Mewaldt, L. R. (1963). California crowned sparrows return from Louisiana. *West. Bird Bander* **38**, 1–4.

Mewaldt, L. R. (1964a). California sparrows return from diplacement to Maryland. *Science* **146**, 941–942.

Mewaldt, L. R. (1964b). Effects of bird removal on a winter population of sparrows. *Bird-Banding* **35**, 184–195.

Mewaldt, L. R. and Rose, R. G. (1960). Orientation of migratory restlessness in the white-crowned sparrow. *Science* **131**, 105–106.

Mewaldt, L. R., Morton, M. L. and Brown, I. L. (1964). Orientation of migratory restlessness in *Zonotrichia. Condor* **66**, 377–417.

Mewaldt, L. R., Cowley, L. T. and Pyong-Oh Won. (1973). California sparrows fail to return from displacement to Korea. *Auk* **90**, 857–861.

Meyer, M. E. (1964). Discriminative basis for astronavigation in birds. *J. Comp. Physiol. Psychol.* **58**, 403–406.

Meyer, M. E. and Lambe, D. R. (1966). Sensitivity of the pigeon to changes in the magnetic field. *Psychon. Sci.* **5**, 349–350.

Michelsen, W. J. (1959). Procedure for studying olfactory discrimination in pigeons. *Science* **130**, 630–631.

Michener, M. C. and Walcott, C. (1966). Navigation of single homing pigeons: aeroplane observations by radio tracking. *Science* **154**, 410–413.

Michener, M. C. and Walcott, C. (1967). Homing of single pigeons—analysis of tracks. *J. Exp. Biol.* **47**, 99–131.

Miselis, R. and Walcott, C. (1970). Locomotor activity rhythms in homing pigeons (*Columba livia*). *Anim. Behav.* **18**, 544–551.

Moore, F. (1977). Geomagnetic disturbance and the orientation of nocturnally migrating birds. *Science* **196**, 682–684.

Montgomery, K. C. and Heinemann, E. G. (1952). Concerning the ability of homing pigeons to discriminate patterns of polarized light. *Science* **116**, 454–456.

Muller, R. E. (1972). Effects of weather on the night time activity of white-throated sparrows. Thesis, Cornell University.

Nisbet, I. C. T. and Drury, W. H. (1968). Short-term effects of weather on bird migration: A field study using multivariate statistics. *Anim. Behav.* **16**, 496–530.

Oehme, H. (1962). Das Auge von Mauersegler, Star und *Amsel. J. Ornithol.* **103**, 187–212.

Orgel, A. R. and Smith, J. C. (1954). Test of the magnetic theory of homing. *Science* **120**, 891–892.

Orr, R. T. (1970). "Animals in Migration." Collier-Macmillan, London.

Oshima, K., Hahn, W. E. and Gorbman, A. (1969a). Olfactory discrimination of natural waters by salmon. *J. Fish. Res. Bd. Can.* **26**, 2111–2121.

Oshima, K., Hahn, W. E. and Gorbman, A. (1969b). Electroencephalographic olfactory response in adult salmon to water traversed in the homing migration. *J. Fish. Res. Bd. Can.* **26**, 2123–2133.

Papi, F. (1976). The olfactory navigation system of homing pigeons. *Verh. Deut. Zool. Ges.* **1976**, 184–205.

Papi, F., Fiore, L., Fiaschi, V. and Benvenuti, S. (1971). The influence of olfactory nerve section on the homing capacity of carrier pigeons. *Monitore Zool. Ital.* (N.S.) **5**, 265–267.

Papi, F., Fiore, L., Fiaschi, V. and Benvenuti, S. (1972). Olfaction and homing in pigeons. *Monitore Zool. Ital.* (N.S.) **6**, 85–95.

Papi, F., Fiaschi, V., Benvenuti, S. and Baldaccini, N. E. (1973). Pigeon homing: outward journey detours influence the initial orientation. *Monitore Zool. Ital.* (N.S.) **7**, 129–133.

Papi, F., Ioalé, P., Fiaschi, V., Benvenuti, S. and Baldaccini, N. E. (1974). Olfactory navigation of pigeons: the effect of treatment with odorous air currents. *J. Comp. Physiol.* **94**, 187–193.

Papi, F., Ioalé, P., Fiaschi, V., Benvenuti, S. and Baldaccini, N. E. (1978). Pigeon homing: cue detection during outward journey and initial orientation. *In* "Animal Migration, Navigation and Homing". (K. Schmidt-Koenig and W. T. Keeton, Eds) pp. 65–77. Proceedings in Life Science. Springer Verlag, Berlin, Heidelberg, New York.

Papi, F., Keeton, W. T., Brown, A. I. and Benvenuti, S. (1979). Do American and Italian Pigeons rely on different homing mechanisms? *J. Comp. Physiol.* **128**, 303–317.

Pennycuick, C. J. (1960). The physical basis of astronavigation in birds: Theoretical considerations. *J. Exp. Biol.* **37**, 573–593.

Perdeck, A. C. (1957). Stichting Vogeltrekstation Texel. Jaarverslag over 1956. *Limosa* **30**, 66–74.

Perdeck, A. C. (1958). Two types of orientation in migrating starlings *Sturnus vulgaris* L., and chaffinches, *Fringilla coelebs* L., as revealed by displacement experiments. *Ardea* **46**, 1–37.

Perdeck, A. C. (1963). Does navigation without visual clues exist in robins? *Ardea* **51**, 91–104.

Perdeck, A. C. (1964). An experiment on the ending of autumn migration in starlings. *Ardea* **52**, 133–139,

Perdeck, A. C. (1967). Orientation of starlings after displacement to Spain. *Ardea* **55**, 194–202.

Pettigrew. J. D. (1978). A role for the avian pecten oculi in orientation to the sun? *In* "Animal Migration. Navigation and Homing". (K. Schmidt-Koenig and W. T. Keeton, Eds) pp. 42–54. Proceedings in Life Sciences. Springer Verlag, Berlin, Heidelberg, New York.

Phillips, D. S. (1969). Effects of olfactory bulb ablation on visual discrimination. *Physiol. Behav.* **5**, 13–15.,

Picton, H. D. (1966). Some response of Drosophila to weak magnetic and electrostatic fields. *Nature* **211**, 303–304.

Pratt, J. G. and Thouless, R. H. (1955). Homing orientation in pigeons in relation to opportunity to observe the sun before release. *J. Exp. Biol.* **32**, 140–157.

Rabøl, J. (1969). Orientation of autumn migrating garden warblers (*Sylvia borin*) after displacement from western Denmark (Blavand) to eastern Sweden (Ottenby). A preliminary experiment. *Dan. Ornithol. Foren. Tidsskr.* **63**, 93–104.

Rabøl, J. (1970). Displacement and phaseshift experiments with night-migrating passerines. *Ornis Scand.* **1**, 27–43.

Rabøl, J. (1972). Displacement experiments with night-migrating passerines (1970). *Z. Tierpsychol.* **30**, 14–25.

Ralph, C. J. and Mewaldt, L. R. (1975). Timing of site fixation upon the wintering grounds in sparrows. *Auk* **92**, 698–705.

Ralph, C. J. and Mewaldt, L. R. (1976). Homing success in wintering sparrows. *Auk* **93**, 1–14.

Rawson, K. S. and Rawson, A. M. (1955). The orientation of homing pigeons in relation to change in sun declination. *J. Ornithol.* **96**, 168–172.

Reille, A. (1968). Essai de mise en évidence d'une sensibilité du pigeon au champ magnétique à l'aide d'un conditionnement nociceptif. *J. Physiol. Paris* **60**, 85–92.

Rendahl, H. (1965). Die Zugverhältnisse der schwedischen Ringeltauben (*Columba palumbus* L.) und Hohltauben (*Columba oenas* L.). *Ark. Zool.* **18**, 221–266.

Richardson, W. J. (1971). Spring migration and weather in eastern Canada: A radar study. *Amer. Birds* **25**, 684–690.

Richardson, W. J. (1972). Autumn migration and weather in eastern Canada: A radar study. *Amer. Birds* **26**. 10–17.

Richardson, W. J. (1974). Bird migration over southeastern Canada, the western Atlantic, and Puerto Rico: a radar study. Ph. D. Thesis, Cornell University, Ithaca, N.Y.

Riper, W. van and Kalmbach, E. R. (1952). Homing not hindered by wing magnets. *Science* **115**, 577–578.

Roadcap, R. (1962). Translocations of white-crowned and golden-crowned sparrows. *West. Bird Bander* **37**, 55–57.

Rüpell, W. (1936). Heimfindeversuche mit Staren und Schwalben 1935. *J. Ornithol.* **84**, 180–196.

Salmonsen, F. (1969). "Vogelzug." BLV Verlagsgesellschaft, München.

Santschi, F. (1911). Observations et remarques critiques sur le mecanisme de l'orientation chez les fourmis. *Rev. Suisse Zoologie* **19**, 303–338.

Sauer, F. (1956). Zugorientierung einer Mönchsgrasmücke (*Sylvia a. atricapilla*, L.) unter künstlichem Sternhimmel. *Naturwiss.* **43**, 231–232. .

Sauer, F. (1957). Die Sternorientierung nächtlich ziehender Grasmücken (*Sylvia atricapilla, borin und curruca*). *Z. Tierpsychol.* **14**, 29–70.

Sauer, E. G. F. (1961). Further studies on the stellar orientation of nocturnally migrating birds. *Psychol. Forschung* **26**, 224–244.

Sauer, E. G. F. (1963). Migration habits of Golden Plovers. *Proc. XIII. Int. Ornithol. Congr.* Ithaca, N.Y., 454–467.

Sauer, F. and Sauer, E. (1955). Zur Frage der nächtlichen Zugorientierung von Grasmücken. *Rev. Suisse Zool.* **62**, 250–259.

Sauer, E. G. F. and Sauer, E. M. (1960). Star migration of nocturnal migrating birds. The 1958 planetarium experiments. *Cold Spring Harb. Symp. Quant. Biol.* **25**, 463–473.

Schlichte, H. J. (1971). Untersuchungen über die Bedeutung optischer Parameter für das Heimkehrverhalten der Brieftaube. Dissertation, Universität Göttingen.

Schlichte, H. J. (1973). Untersuchungen über die Bedeutung optischer Parameter für das Heimkehrverhalten der Brieftaube. *Z. Tierpsychol.* **32**, 257–280.

Schlichte, H. J. and Schmidt-Koenig, K. (1971). Zum Heimfindevermögen der Brieftaube bei erschwerter optischer Wahrnehmung. *Naturwiss.* **58**, 329–330.

Schmidt-Koenig, K. (1958). Experimentelle Einflussnahme auf die 24-Stunden-Periodik bei Brieftauben und deren Auswirkungen unter besonderer Berücksichtigung des Heimfindevermögens. *Z. Tierpsychol.* **15**, 301–331.

Schmidt-Koenig, K. (1960). Internal clocks and homing. *Cold Spring Harb. Symp. Quant. Biol.* **25**, 389–393.

Schmidt-Koenig, K. (1961). Die Sonne als Kompass im Heim-Orientierungssystem der Brieftauben. *Z. Tierpsychol.* **68**, 221–244.

Schmidt-Koenig, K. (1963a). Sun compass orientation of pigeons upon displacement north of the arctic circle. *Biol. Bull.* **127**, 154–158.

Schmidt-Koenig, K. (1963b). Sun compass orientation of pigeons upon equatorial and trans-equatorial displacement. *Biol. Bull.* **124**, 311–321.

Schmidt-Koenig, K. (1963c). On the role of the loft, the distance and site or release in pigeon homing (the "cross loft experiment"). *Biol. Bull.* **125**, 154–164.

Schmidt-Koenig, K. (1963d). Neuere Aspekte über die Orientierungsleistungen von Brieftauben. *Ergeb. Biol.* **26**, 286–297.

Schmidt-Koenig, K. (1964). Über den zeitlichen Ablauf der Anfangsorientierung bei Brieftauben (Kurzfassung). *Verh. Deut. Zool. Ges.* **1964**, 407–411.

Schmidt-Koenig, K. (1965). Current problems in bird orientation. *In* "Advances in Study Behavior". (D. S. Lehrman, R. A. Hinde, E. Shaw, Eds) Vol. 1. pp. 217–278. Academic Press, New York and London.

Schmidt-Koenig, K. (1966). Über die Entfernung als Parameter bei der Anfangsorientierung der Brieftaube. *Z. Vergl. Physiol.* **52**, 33–55.

Schmidt-Koenig, K. (1969). Weitere Versuche, durch Verstellen der inneren Uhr in den Heimkehrprozess der Brieftaube einzugreifen. *Verh. Deut. Zool. Ges.* **33**, 200–205.

Schmidt, K. (1970a). Ein Versuch. theoretisch mögliche Navigationsverfahren von Vögeln zu klassifizieren und relevante sinnesphysiologische Probleme zu umreissen. *Verh. Deut. Zool. Ges.* **1970**, 243–245.

Schmidt-Koenig, K. (1970b). Entfernung und Heimkehrverhalten der Brieftaube. *Z. Vergl. Physiol.* **68**, 39–48.

Schmidt-Koenig, K. (1970c). "Bird Navigation" (Kritik an G. V. T. Matthews), 2nd ed. Cambridge Univ. Press, Cambridge: (1968). *Z. Tierpsychol.* **27**, 118–121.

Schmidt-Koenig, K. (1971). Remarks in a panel discussion: Unconventional theories of orientation. *Ann. N.Y. Acad. Sci.* **188**, 338–339.

Schmidt-Koenig, K. (1972). New experiments on the effect of clock shifts on homing pigeons. *In* "Animal Orientation and Navigation". (S. R. Galler, K. Schmidt-Koenig, G. J. Jacobs and R. E. Belleville, Eds) pp. 275–285. NASA SP-262 US Govt. Printing Office, Washington D.C.

Schmidt-Koenig, K. (1975). "Migration and Homing in Animals." Springer Verlag, Berlin, Heidelberg, New York.

Schmidt-Koenig, K. (1976). Direction of the preceding release and initial orientation of homing pigeons. *Auk* **93**, 669–674.

Schmidt-Koenig, K. (1978). On the role of olfactory cues in pigeon homing. *Proc. 17. Int. Ornithol. Congr.*

Schmidt-Koenig, K. and Keeton W. T. (1977). Sun compass utilization by pigeons wearing frosted contact lenses. *Auk* **94**, 143–145.

Schmidt-Koenig, K. and Keeton W. T. (1978). Eds "Animal Migration, Navigation and Homing." Series: Proceedings in Life Sciences, Springer Verlag, Berlin, Heidelberg, New York.

Schmidt-Koenig, K. and Phillips, J. B. (1978). Local anethesia of the olfactory membrane and homing in pigeons. *In* "Animal Migration, Navigation and Homing". (K. Schmidt-Koenig and W. T. Keeton, Eds) pp.119–124. Proceedings in Life Sciences. Springer Verlag, Berlin, Heidelberg, New York.

Schmidt-Koenig, K., and Schlichte, H. J. (1972). Homing in pigeons with impaired vision. *Proc. Nat. Acad. Sci. USA* **69**, 2446–2447.

Schmidt-Koenig, K. and Walcott, C. (1973). Flugwege und Verblieb von Brieftauben mit getrübten Haftschalen. *Naturwiss.* **60**, 108–109.

Schmidt-Koenig, K. and Walcott, C. (1978). Tracks of pigeons homing with frosted lenses. *Anim. Behav.* **26**, 480–486.

Schreiber, B., Gualtierotti, T. and Mainardi, D. (1962). Some problems of cerebellar physiology in migratory and sedentary birds. *Anim. Behav.* **10**, 42–47.

Schüz, E. (1971). "Grundriss der Vogelzugskunde." Parey Verlag, Berlin and Hamburg.

Serventy, D. L. (1953). Movements of pelagic sea birds in the Indo-Pacific region. *Proc. VII. Pacif. Sci. Congr.* **4**, 394–407.

Serventy, D. L. (1967). Aspects of the population ecology of *Puffinus tenuirostris. Proc. XIV. Ornithol. Congr.* Oxford, 165–190.

Shallenberger, R. J. (1973). Breeding biology, homing behaviour, and communication patterns of the wedge-tailed shearwater, *Puffinus pacificus.* Ph. D. Thesis, University of California, Los Angeles.

Shibuya, T. and Tucker, D. (1967). Single unit response of olfactory receptors in vulture. *In* "Olfaction and Taste". (T. Hayashi, Ed.) Vol. 2, pp. 219–233. Pergamon Press, Oxford.

Shumake, S. A., Smith, J. C. and Tucker D. (1969). Olfactory intensity-difference thresholds in the pigeon. *J. Comp. Physiol. Psychol.* **67**, 64–69.

Shumakov, M. E. (1965. [Preliminary results of the investigation of migrational orientation of passerine birds by the round-cage method.] (In Russian.) *In* "Bionika." Moscow, pp. 371–378.

Shumhakov, M. E. (1967). [An investigation of the migratory orientation of passerine birds.] Vestn. Leningrad. Uni., Biol. Ser. (in Russian) **1967** (**3**), 106–118.

Sieck, M. H. and Wenzel, B. M. (1969). Electrical activity of the olfactory bulb of the pigeon. *Electroenceph Clin. Neurophysiol.* **26**, 62–69.

Slepian, J. (1948). Physical basis of bird navigation. *J. Appl. Phys.* **19**, 306.

Snyder, R. L. and Cheney, C. D. (1975). Homing performance of anosmic pigeons. *Bull. Psychonom. Soc.* **6**, 592–594.

Sobol, E. D. (1930). Orienting ability of carrier pigeons with injured labyrinths. Mil.-Med. Zh. USSR 1, 75. *Biol. Abstr.* **8**, 15425.

Sokolov, L. V. (1970. [Experimental study using Kramer cages, of the orientational abilities of birds with various types of movement.] *In* "Ornithology in the USSR". Proc. 7th Baltic Ornithol. Conf., 1970, Vol. 2, pp. 129–133 (in Russian) Ashkhabad.

Sonnberg, A. (1972). Dressurversuche mit Brieftauben: Eine Untersuchung über die Fähigkeit der Brieftaube (*Columba livia*), Gegenstände zu erkennen und zur Orientierung zu verwenden. Dissertation, Universität Göttingen.

Southern, W. E. (1959). Homing of purple martins. *Wilson Bull.* **71**, 254–261.

Southern, W. E. (1968). Experiments on the homing ability of purple martins. *The Living Bird* **7**, 71–84.

Southern, W. E. (1969). Orientation behaviour of ring-billed gull chicks and fledglings. *Condor* **71**, 418–425.

Southern, W. E. (1978). Orientation responses of ring-billed gull chicks: a re-evaluation. *In* "Animal Migration, Navigation and Homing". (K. Schmidt-Koenig and W. T. Keeton, Eds) pp. 311–317. Proceedings in Life Sciences. Springer Verlag, Berlin, Heidelberg, New York.

Stager, K. E. (1964). The role of olfaction in food location by the turkey vulture (*Cathartes aura*). Los Angeles County Mus. Contrib. Sci. No. 81.

St. Paul, U. von (1956). Compass directional training of western meadow-larks (*Sturnella reglecta*). *Auk* **73**, 203–210.

St. Paul, U. von (1962). Das Nachtfliegen von Brieftauben. *J. Ornithol.* **103**, 337–343.

Storr, G. M. (1958). Migration routes of the arctic tern. *Emu* **58**, 59-62.

Strong, R. M. (1911). On the olfactory organs and the sense of smell in birds. *J. Morphol.* **22**, 619–661.

Thorpe, W. H. and Wilkinson, D. H. (1946). Ising's theory of bird orientation. *Nature* **158**, 903.

Tucker, D. (1965). Electrophysical evidence for olfactory function in birds. *Nature, Lond.* **207**, 34–36.

Varian, R. H. (1948). Remarks on: "A preliminary study of a physical basis of bird navigation". *J. Appl. Phys.* **19**, 306–307.

Vries, H. de (1948). Die Reizschwelle der Sinnesorgane als physikalisches Problem. *Experientia* **4**, 205–213.

Wagner, G. (1968) Topographisch bedingte zweigipflige und schiefe Kreisverteilungen bei der Anfangsorientierung verfrachteter Brieftauben. *Rev. Suisse Zool.* **75**, 682–690.

Wagner, G. (1970). Verfolgung von Brieftauben im Helikopter. *Rev. Suisse Zool.* **77**, 39–60.

Wagner, G. (1972). Topography and pigeon orientation. *In* "Animal Orientation and Navigation". (S. R. Galler, K. Schmidt-Koenig, G. J. Jacobs and R. E. Belleville, Eds) pp. 249–273. NASA SP-262 US Govt. Printing Office, Washington D.C.

Wagner, G. (1974). Verfolgung von Brieftauben im Helikopter II. *Rev. Suisse Zool.* **80**, 727–750.

Wagner, G. (1978). Homing pigeons flight over and under low stratus. *In* "Animal Migration, Navigation and Homing". (K. Schmidt-Koenig and W. T. Keeton, Eds.) Proceedings in Life Sciences. Springer Verlag, Berlin, Heidelberg, New York.

Walcott, C. (1972). The navigation of homing pigeons: do they use sun navigation? *In* "Animal Orientation and Navigation". (S. R. Galler, K. Schmidt-Koenig, G. J. Jacobs and R. E. Belleville, Eds) pp. 283–292. NASA SP-262 US Govt. Printing Office, Washington D.C.

Walcott, C. (1977). Magnetic fields and the orientation of homing pigeons under sun. *J. Exp. Biol.* **70**, 105–123.

Walcott, C. (1978). Anomalies in the earth's magnetic field increase the scatter of pigeon's vanishing bearings. *In* "Animal Migration, Navigation and Homing". (K. Schmidt-Koenig and W. T. Keeton, Eds (pp. 143–151. Proceedings in Life Sciences. Springer Verlag, Berlin, Heidelberg, New York.

Walcott, C. and Green, R. P. (1974). Orientation of homing pigeons altered by a change in the direction of an applied magnetic field. *Science* **184**, 180–182.

Walcott, C. and Michener, M. C. (1971). Sun navigation in homing pigeons—attempts to shift sun co-ordinates. *J. Exp. Biol.* **54**, 291–316.

Walcott, C. and Schmidt-Koenig, K. (1973). The effect on pigeon homing of anesthesia during displacement. *Auk* **90**, 281–286.

Wallraff, H. G. (1959). Örtlich and zeitlich bedingte Variabilität des Heimkehrverhaltens von Brieftauben. *Z. Tierpsychol.* **16**, 513–544.

Wallraff, H. G. (1960a). Uber Zusammenhänge des Heimkehrverhaltens von Brieftauben mit meteorologischen und geophysikalischen Faktoren. *Z. Tierpsychol.* **17**, 82–113.

Wallraff, H. G. (1960b). Können Grasmücken mit Hilfe des Sternenhimmels navigieren? *Z. Tierpsychol.* **17**, 165-177.

Wallraff, H. G. (1960c). Does celestian navigation exist in animals? *Cold Spring Harb. Symp. Quant. Biol.* **25**, 451–461.

Wallraff, H. G. (1965). Über das Heimfindevermögen von Brieftauben mit durchtrennten Bogengängen. *Z. Vergl. Physiol.* **50**, 313–330.

Wallraff, H. G. (1966a). Versuche zur Frage der gerichteten Nachtzug-Aktivität von gekäfigten Singvögeln. *Verh. Deut. Zool. Ges. Jena* **1965**, 338–355.

Wallraff, H. G. (1966b). Über die Anfangsorientierung von Brieftauben unter geschlossener Wolkendecke. *J. Ornithol.* **107**, 326-336.

Wallraff, H. G. (1966c). Über die Heimfindeleistung von Brieftauben nach Haltung in verschiedenartig abgeschirmten Volieren. *Z. Vergl. Physiol.* **52**, 215–259.

Wallraff, H. G. (1967). The present status of our knowledge about pigeon homing. *Proc. XIV. Int. Ornithol. Congr.*, Oxford, 331–358.

Wallraff, H. G. (1969). Über das Orientierungsvermögen von Vögeln unter natürlichen und künstlichen Sternmustern. Dressurversuche mit Stockenten. *Ver. Deut. Zool Ges.* Innsbruck, 1968, 348–357.

Wallraff, H. G. (1970a). Weitere Volierenversuche mit Brieftauben: Wahrscheinlicher Einflus dynmischer Faktoren der Atmosphäre auf die Orientierung. *Z. Vergl. Physiol.* **68**, 182–201.

Wallraff, H. G. (1970b). Über die Flugrichtungen verfrachteter Brieftauben in Abhängigkeit vom Heimatort und vom Ort der Freilassung. *Z. Tierpsychol.* **27**, 303–351.

Wallraff, H. G. (1971). Kurzzeitige Schwankungen bei der Richtungswahl abfliegender Brieftauben. *J. Ornithol.* **112**, 396–410.

Wallraff, H. G. (1972a). Nichtvisuelle Orientierung zugunruhiger Rotkelchen (*Erithacus rubecula*). *Z. Tierpsychol.* **30**, 374–383.

Wallraff, H. G. (1972b). Homing of pigeons after extirpation of their cochlae and legenae. *Nature Lond.* **236**, 223–224.

Wallraff, H. G. (1974a). "Das Navigationssytem der Vögel." Oldenbourg, München, Wien.

Wallraff, H. G. (1974b). The effect of directional experience on initial orientation in pigeons. *Auk* **91**, 24–34.

Wallraff, H. G. and Graue, L. C. (1973). Orientation of pigeons after transatlantic displacement. *Behaviour* **44**, 1–35.

Walls, G. L. (1963). The Vertebrate Eye and its Adaptive Radiation. New York and London.

Watson, G. S. (1961). Goodness-of-fit tests on a circle. *Biometrika* **48**, 109–114.

Watson, G. S. (1962). Goodness-of-fit test on a circle II. *Biometrika* **49**, 57–63.

Watson, G. S. and Williams, E. J. (1956). On the construction of significance tests on the circle and the sphere. *Biometrika* **43**, 344–352.

Wenzel, B. M. (1967). Olfactory perception in birds. *In* "Olfaction and Taste" (T. Hayashi, Ed.) Vol 2, pp. 203–217. Pergamon Press, Oxford.

Wenzel, B. M. (1971a). Olfaction in birds. *In* "Handbook of Sensory Physiology. (L. B. Beidler, Ed.) Vol. IV, pp 432–448. Springer Verlag, Berlin, Heidelberg, New York.

Wenzel, B. M. (1971b). Olfactory sensation in the Kiwi and other birds. *Ann. N.Y. Acad. Sci.* **188**, 183–193.

Wenzel, B. M. and Salzman, A. (1968). Olfactory bulb ablation or nerve section and behaviour of pigeons in nonolfactory learning. *Exp. Neurobiol.* **22**, 472–479.

Wenzel, B. M. and Sieck, M. H. (1972). Olfactory perception and bulbar electrical activity in several avian species. *Physiol. Behav.* **9**, 287–294.

Whiten, A. (1972). Operant study of sun altitude and pigeon navigation. *Nature, Lond.* **237**, 405–406.

Whitten, A. J. (1971). A new behavioural method for further determination of olfaction in Mallard (*Anasplatyrhynchos.*) *J. Biol. Educ.* **5**, 291–294.

Williams, T. C., Williams, J. M., Teal, J. M. and Kanwisher, J. W. (1972). Tracking radar studies of bird migration. *In* "Animal Orientation and Navigation". (S. R. Galler, K. Schmidt-Koenig, G. J. Jacobs and R. E. Belleville, Eds) pp. 115–128. NASA SP-262 US Govt. Printing Office, Washington D.C.

Williams, J. M., Williams, T. C. and Ireland, L. C. (1974). Bird Migration over the North Atlantic. *In* "The biological aspects of the bird/aircraft collision problem. (S. Gauthreaux, Ed.) Air Force Office of Scientific Research, Clemson N.C., USA.

Williams, T. C., Williams, J. M., Ireland, L. C. and Teal, J. M. (1977). Autumnal bird migration over the western North Atlantic Ocean. *Amer. Birds* **31**. 251–267.

Williams, T. C. and Williams, J. M. (1978). Orientation of transatlantic migrants. *In* "Animal Migration, Navigation and Homing". (K. Schmidt-Koenig and W. T. Keeton, Eds) pp. 239–251. Proceedings in Life Sciences. Springer Verlag, Berlin, Heidelberg, New York.

Wiltschko, R. and Wiltschko, W. (1978). Evidence for the use of magnetic outward-journey information in homing pigeons. *Naturwiss.* **65**, 112–113.

Wiltschko, R., Wiltschko, W. and Keeton, W. T. (1978). Effect of outward journey in an altered magnetic field in young homing pigeons. *In* "Animal Migration, Navigation and Homing". (K. Schmidt-Koenig and W. T. Keeton, Eds) pp. 152–161. Proceedings in Life Sciences. Springer Verlag, Berlin, Heidelberg, New York.

Wiltschko, W. (1968). Über den Einfluß statischer Magnetfelder auf die Zugorientierung der Rotkehlchen (*Erithacus rubecula*). *Z. Tierpsychol.* **25**, 537–558.

Wiltschko, W. (1972). The influence of magnetic total intensity and inclination on directions preferred by migrating European robins (*Erithacus rubecula*). *In* "Animal Orientation and Navigation". (S. R. Galler, K. Schmidt-Koenig, G. J. Jacobs and R. E. Belleville, (Eds) pp. 569–578. NASA SP-262 US Govt. Printing Office, Washington D.C.

Wiltschko, W. (1973). Kompaßsysteme in der Orientierung von Zugvögeln. Akad. Wiss. Lit. Mainz. Reihe Inf. Org. II., Steiner Verl. Wiesbaden.

Wiltschko, W. (1978). Further analysis of the magnetic compass of migratory birds. *In* "Animal Migration, Navigation and Homing". K. Schmidt-Koenig and W. T. Keeton, Eds) pp. 302–310. Proceedings in Life Sciences. Springer Verlag, Berlin, Heidelberg, New York.

Wiltschko, W. and Wiltschko, R. (1972). The magnetic compass of European robins. *Science* **176**, 62-64.

Wiltschko, W. and Wiltschko, R. (1975a). The interaction of stars and magnetic field in the orientation system of night migrating birds. Part I. Autumn experiments with European warblers (Gen. *Sylvia*). *Z. Tierpsychol.* **37**, 337–355.

Wiltschko, W. and Wiltschko R. (1975b). The interaction of stars and magnetic field in the orientation system of night migrating birds. Part II. Spring experiments with European robins (*Erithacus rubecula*). *Z. Tierpsychol.* **39**, 265–282. ,

Wiltschko, W. and Wiltschko, R. (1976a). Interrelation of magnetic compass and star orientation in night-migrating birds. *J. Comp. Physiol.* **109**, 91–99.

Wiltschko, W. and Wiltschko, R. (1976b). Die Bedeutung des Magnetkompasses für die Orientierung der Vögel. *J. Ornithol.* **117**, 362–387.

Wiltschko, W., Wiltschko, R. and Keeton, W. T. (1976). Effects of a "permanent" clock-shift on the orientation of young homing pigeons. *Behav. Ecol. Sociobiol.* **1**, 229–243.

Wingstrand, K. G. and Munk, O. (1965). The pecten oculi of the pigeon with particular regard to its function. *Biol. Skr. Dan. Vid. Selsk.* **14**, 1-64.

Windsor, D. M. (1975). Regional expression of directional preferences by experienced homing pigeons. *Anim. Behav.* 23, 335–343.

Yeagley, H. L. (1947). A preliminary study of a physical basis of bird navigation. *J. Appl. Phys.* **18**, 1035–1063.

Yeagley, H. L. (1951). A preliminary study of a physical basis of bird navigation II. *J. Appl. Phys.* **22**, 746–760.

Yodlowski, M. L., Kreithen, M. L. and Keeton, W. T. (1977). Detection of atmospheric infrasound by homing pigeons. *Nature* **265**, 725–726.

Index